W0043877

SPRINGER TRACTS IN MODERN PHYSICS

Ergebnisse der exakten Naturwissenschaften

Volume 53

Editor: G. Höhler

Editorial Board: P. Falk-Vairant S. Flügge J. Hamilton F. Hund H. Lehmann E.A. Niekisch W. Paul

Springer-Verlag Berlin Heidelberg GmbH 1970

Manuscripts for publication should be addressed to:

G. HÖHLER, Institut für Theoretische Kernphysik der Universität, 75 Karlsruhe, Kaiserstraße 12

Proofs and all correspondence concerning papers in the process of publication should be addressed to:

E. A. NIEKISCH, Kernforschungsanlage Jülich, Institut für Technische Physik, 517 Jülich, Postfach 365

ISBN 978-3-662-15887-6 ISBN 978-3-540-36317-0 (eBook)
DOI 10.1007/978-3-540-36317-0

This work is subject to copyright. All rights are reserved, whether the whole or part of the material is concerned, specifically those of translation, reprinting, re-use of illustrations, broadcasting, reproduction by photocopying machine or similar means, and storage in data banks.

Under § 54 of the German Copyright Law where copies are made for other than private use, a fee is payable to the publisher, the amount of the fee to be determined by agreement with the publisher. © by
Springer-Verlag Berlin Heidelberg 1970
Originally published by Springer-Verlag Berlin Heidelberg New York in 1970
Softcover reprint of the hardcover 1st edition 1970

Library of Congress Catalog Card Number 25-9130.

The use of general descriptive names, trade names, trade marks, etc. in this publication, even if the former are not especially identified, is not be taken as a sign that such names, as understood by the Trade Marks and Merchandise Marks Act, may accordingly be used freely by anyone. Title-No. 4736

Contents

On the S-Matrix Theory of Weak Interactions*

A. O. Barut

It is generally accepted that the theory of weak interactions is not on as firm a ground as the theory of electromagnetic interactions. In fact, the present theory of weak interactions (and I shall confine myself to leptonic weak interactions, μ-decay, β-decay, etc.) is basically the same as Fermi's initial treatment [1], which in turn expresses almost entirely the kinematics of the processes, *lumping* the dynamics into a phenomenological coupling constant G, to be determined experimentally. The post-Fermi developments that fixed the specific form of the kinematical part of the four-fermion coupling are:

(1°) The $(V-\lambda A)$ form of the coupling [2] ($\lambda=1$ for μ-decay),

(2°) Two two-component neutrinos ν and $\bar{\nu}$ [3], and the experimental verification that $\nu \neq \bar{\nu}$,

(3°) Universal Fermi interactions, conserved vector currents [4], and generalization to other members of $SU(3)$ – multiplets [5].

(4°) Evaluation of the λ factor in $(V-\lambda A)$ for β-decay [6].

It is a new step in the development of the theory if the Fermi-form of the matrix elements is *elevated* to an interaction Lagrangian, i.e.

$$\mathscr{L}_{\text{int.}} = \frac{G}{\sqrt{2}} \int \mathrm{d}x\, \bar{\psi}_\mu(x)\, \gamma_\mu(1-\gamma_5)\, \psi_\nu(x)\, \bar{\psi}_e(x)\, \gamma^\mu(1-\gamma_5)\, \psi_\nu(x)\,. \qquad (1)$$

To first order in G, the matrix elements of $\mathscr{L}_{\text{int.}}$ coincide with the "kinematical matrix elements". But if the quantum field theory with the coupling (1) ($G=\text{const.}$) is taken seriously, one has the following problems:

(a) The theory is non-renormalizable, all higher order terms are infinite;

(b) even the lowest order calculation gives incorrect results at high energies (the cross-section becomes larger than the value allowed by unitarity);

(c) the electromagnetic (radiative) corrections are infinite (in the case of semi-leptonic interactions).

---

* Lecture given at the Summer Institute on Weak Interactions, University of Karlsruhe, July 1969.

Under these circumstances we can say that we do not have a dynamical theory of weak interactions at all. There are those who say that the basic form of the interaction, Eq. (1), is correct, but we do not know how to calculate with this interaction. This has to be demonstrated. Others say that Eq. (1) must be used as a phenomenological Lagrangian only in first order. This is also not satisfactory, because as we noted above, even the first order result is not consistent at high energies. Finally, there are attempts to give a structure to the form of the interaction, in the form of intermediate bosons or otherwise [7], but keeping G constant in Eq. (1). Theories with intermediate bosons are also non-renormalizable. (There are clearly many other ingenious but complex procedures and proposals for a quantum field theory of weak interactions [8].) It is also generally agreed that the outstanding fundamental problem in weak interactions theory is finding a way of dealing with higher order effects.

We wish to continue here the original kinematical approach of Fermi, and taking the form (1) not as a Lagrangian but as the kinematical basis function of the amplitude itself, and wish to show that under very general conditions G must be a decreasing function of the energy and momentum transfer, at least as fast as $1/(s \log s)$. In other words we shall view G as the scalar amplitude for the process. In fact, this is the leptonic counterpart of the dispersion treatment of semi-leptonic weak interactions [9], where one writes a general form of the amplitude with a number of form factors.

The starting points of our considerations are then

(i) the general form of the amplitudes as derived from Lorentz invariance, time reversal and $e-\mu$ universality, and other possible symmetries,

(ii) the unitarity condition,

(iii) some analyticity properties.

To justify the point (iii) I shall argue that inasmuch as the weak interactions are short-ranged and the physical region analyticity of the scattering amplitude follows from the causality requirement and short range nature of the interactions, we should be able to use the analyticity properties of the amplitude with at least as much confidence as in strong interactions. In fact, an S-matrix approach even works for electromagnetic interactions [10].

In this lecture I consider the simplest leptonic interaction [11], $e\nu \rightarrow e\nu$. The process is governed by a single amplitude which can be chosen to be of $(V-A)$ type. The elastic unitarity condition gives after some manipulation [11] a simple relation for the scalar amplitude $G(s,t)$:

$$\operatorname{Im} G(s,t) = \frac{\sqrt{2}}{2}(2\pi)^{-4}\frac{(s-m_e^2)^2}{s}\int d\Omega\, G(s,t')\, G^*(s,t'') . \tag{2}$$

The only difference as compared to the $\pi-\pi$ scattering is the new kinematical factor $(s-m_e^2)^2/s$, which, we shall see, has rather important consequences.

The crossed channels are $e\bar{e}\to\bar{\nu}\nu$ and $e\bar{\nu}\to e\bar{\nu}$; and $G(s,t)$ will describe all these three channels in their respective physical regions [11].

We must actually consider the coupled channels consisting besides $e\nu\to e\nu$, also $e\nu\to\mu\nu$ and $\mu\nu\to\mu\nu$ channels, with two types of neutrinos ν_e, ν_μ and with due account of separate conservation of the leptonic numbers L_e and L_μ. For some of the general conclusions that we want to make, it is sufficient, however, to consider a single channel problem.

Many particle intermediate states in the unitarity will give singularities that lie far away. Furthermore, and this is more important, the discontinuities of these singularities will be proportional to $G^2\alpha$, G^4, etc. ..., where the factor α comes from lowest order electromagnetic interactions (radiative corrections). Because we are not aiming at a closed complete theory of weak interactions, but one which is approximate, but consistent and does not violate unitarity, we neglect higher order intermediate states in the unitarity. Thus, to a good approximation, the weak interactions provide us with an example of an amplitude that satisfies elastic unitarity all the way, just like the potential scattering.

The unitarity condition, in terms of the partial wave amplitudes, from (2), is given by

$$\operatorname{Im} G(s,l)=\frac{1}{\sqrt{2}\pi}\,\frac{(s-m_e^2)^2}{s}\,|G(s,l)|^2\,,\qquad s>m_e^2\,,\tag{3}$$

so that the partial wave amplitudes are bounded by

$$|G(s,l)|\leqq\frac{\sqrt{2}\pi\, s}{(s-m_e^2)^2}\to\frac{\sqrt{2}\pi}{s}\,.\tag{4}$$

It is interesting to combine this limit with the so-called analyticity bound to obtain the analog of the Froissart-bound [12] for the total cross-section. If we have an analyticity domain in the z-plane, with a cut at $z_{0_{s\to\infty}}=1+2t/s$ and boundedness of the amplitude $G(s,t)$ by a polynomial in s, we find

$$|G(s,t=0)|\leqq\text{const.}\,(\log s)^2\,.\tag{5}$$

Hence

$$\sigma_{\text{tot}}\leqq\text{const.}\,\frac{(\log s)^2}{s}\,.\tag{6}$$

Recall that the corresponding limit for $\pi\pi$ is: $\sigma_{\text{tot}}\leqq\text{const.}\,(\log s)^2$. Thus weak interaction cross-sections must decrease at high energies. We can obtain a more precise limit on the amplitudes as follows:

The inverse partial wave amplitude $G^{-1}(s, l)$ satisfies

$$\operatorname{Im} G^{-1}(s, l) = -\frac{1}{\sqrt{2\pi}} \frac{(s-m_e^2)^2}{s}. \tag{7}$$

Hence we need two subtractions in the dispersion relation for G^{-1}:

$$G^{-1}(s, l) = A + B(s-s_0) + \frac{\lambda}{\pi}(s-s_0)^2 \int_{m_e^2}^{\infty} \frac{(s'-m_e^2)^2}{s'(s'-s_0)^2(s'-s)} ds'. \tag{8}$$

We may choose $s_0 = m_e^2$, then

$$G^{-1}(s, l) = A + B(s-m_e^2) - \frac{\lambda}{\pi} \frac{(s-m_e^2)^2}{s} \log \frac{m_e^2 - s}{m_e^2}.$$

Because $G(s, l)$ has also a left-hand cut, $A(s, l)$ and $B(s, l)$ in the above formulas, are arbitrary functions with a left-hand cut only, they are meromorphic for $s > 0$. Thus, as $s \to \infty$ $\operatorname{Im} G^{-1} \to -\frac{1}{\sqrt{2\pi}} s$, and $\operatorname{Re} G^{-1} \to A + Bs - \frac{\lambda}{\pi} s \log \frac{s}{m_e^2}$. There are then two cases: Either $(A+Bs) \to s^\alpha$, $\alpha > 1$ in which case $\operatorname{Re} G^{-1} \to s^\alpha$, or $(A+Bs) \to s^\beta$, $\beta < 1$, in which case $\operatorname{Re} G^{-1} \to s \log s$. Thus, either

$$G(s, l) \to \frac{s^\alpha}{s^{2\alpha}+s^2} - i \frac{s}{s^{2\alpha}+s^2}, \qquad \alpha > 1 \tag{9}$$

or

$$G(s, l) \to \frac{s \log s}{(s \log s)^2 + s^2} - i \frac{s}{(s \log s)^2 + s^2}.$$

Consequently

$$|G(s, l)| \leq \frac{1}{s \log s},$$

a limit which is a bit better than the unitarity limit.

In order to obtain a solution of the dispersion relations we must take the symmetry property of $G(s, t)$ and the left-hand cut into account. The iterated solution of the N/D-equations are known if the left-hand cut were approximated by a pole, as in the Zachariasen model [13]. This would be in agreement with the asymptotic values (9) which is derived here (but assumed in the Zachariasen model). However, this approximation does not take into account the symmetry of the problem. Because of the asymptotic $1/s$ behavior of the amplitude G we do not need any subtraction in the dispersion relations for G, and there is the very remarkable possibility that the value of the weak-interaction coupling constant may be calculated (in the elastic unitarity approximation to order α).

References

1. *Fermi, E.*: Nuovo Cimento **11**, 1 (1934); – Z. Physik **88**, 161 (1934).
2. *Sudarshan, E. C. G., Marshak, R. E.*: Proceedings of Padua Conference on Mesons and Recently Discovered Particles (1957). – *Gell-Mann, M., Feynman, R. P.*: Phys. Rev. **109**, 193 (1958).
3. *Lee, T. D., Yang, C. N.*: Phys. Rev. **104**, 254 (1956); **105**, 1671 (1957).
   Salam, Abdus: Nuovo Cimento **5**, 29 (1957).
   Landau, L. D.: JETP **32**, 405 and 407 (1957).
4. *Gell-Mann, M., Feynman, R. P.*: Phys. Rev. **109**, 193 (1958).
   Gershtein, S. S., Zeldovitch, I. B.: Soviet Phys. JETP **2**, 576 (1956).
5. *Cabibbo, N.*: Phys. Rev. Letters **10**, 531 (1963).
6. *Adler, S. L.*: Phys. Rev. Letters **14**, 1051 (1965); – Phys. Rev. **140**, B 736 (1965).
   Weissberger, W. I.: Phys. Rev. Letters **14**, 1047 (1965); – Phys. Rev. **143**, B 1302 (1966).
7. *Lee. T. D., Yang, C. N.*: Phys. Rev. **108**, 1611 (1957).
   Pietschmann, H.: These Proceedings.
8. Cf. S. B. Treiman's Rapporteur's Talk, Proceedings of International High Energy Physics Conference, Vienna, 1968.
9. *Goldberger, M. L., Treiman, S. B.*: Phys. Rev. **111**, 354 (1958).
10. *Barut, A. O.*: The theory of the scattering matrix, Ch. XIII. New York: Macmillan 1967.
11. For general kinematical results, semileptonic reactions, helicity amplitudes, discrete symmetries, etc. ... we refer to *Barut, A. O., Baran, A.*: Trieste preprint 1970.
12. *Froissart, M.*: Phys. Rev. **123**, 1053 (1961).
13. *Zachariasen, F.*: Phys. Rev. **121**, 1851 (1961).
    Thirring, W.: Phys. Rev. **126**, 1209 (1962).

Prof. Dr. *A. O. Barut*
University of Colorado
Boulder, Colorado 80302/USA

Weak Interactions in Nuclear Physics*

H. PRIMAKOFF

Contents

I. Introduction

In the present lectures we shall discuss those aspects of the weak interactions which can be observed most easily in processes where the hadrons involved are nuclei rather than nucleons or nucleon-isobars or mesons; this relative ease of observation arises essentially from the fact that hadrons with baryon number greater then one have a much richer spectrum of masses, spins, charges, isospins and lifetimes than hadrons with baryon number one or zero. In particular, we shall concentrate our attention on the semileptonic weak processes in nuclei and treat in some detail:

(a) "Meson-exchange" and/or "bound nucleon-isobar" effects in nuclear beta decay and nuclear muon capture – quantitative inadequacy of the conventional neutron-proton nuclear model.

(b) Nuclear two-neutrino and no-neutrino double-beta decay – lepton conservation.

(c) "Second-class" hadronic weak currents in nuclei – transformation properties of hadronic weak currents under time reversal and charge symmetry.

---

* Lectures given at the Summer Institute on Weak Interactions, University of Karlsruhe, July 1969.

(d) Goldberger-Treiman relation, Adler-Weisberger sum rule, and Weinberg-Tomozawa formula in nuclei – pion-pole dominance of the axial-current divergence form factor in nuclear beta decay.

On the other hand, we shall give only a very brief treatment of the problem of nonleptonic weak processes in nuclei; these processes give rise to space-inversion noninvariant weak nuclear forces which produce small but observable "wrong-parity" admixtures in the various nuclear states.

II a) Semileptonic Weak Processes in Nuclei — "Meson-Exchange" and/or "Bound Nucleon-Isobar" Effects

The decay rate and energy release in $H^3 \to He^3 + e^- + \nu_e$, together with the CVC-based assumption that [1]

$$F_V(q^2 = 0; H^3 \to He^3) = 1 \tag{1}$$

yield

$$|F_A(q^2 = 0; H^3 \to He^3)| = 1.21 \pm 0.01 \tag{2}$$

where the $F_V(q^2; N_i \to N_f)$ and $F_A(q^2; N_i \to N_f)$ are vector and axial form factors in $N_i \to N_f + e + \nu_e$ and $q^2 = (p_{N_f} - p_{N_i})^2$. On the other hand, the conventional neutron-proton nuclear model, which assumes that nuclei consist only of nucleons and that these nucleons are characterized by the same weak (and electromagnetic) form factors as free nucleons, predicts

$$|F_A(q^2 = 0; H^3 \to He^3)| = \left\{ \left(\frac{J}{J+1}\right) \sum_{M_f = -J}^{J} \left| \langle \Psi_{M_f}(He^3) | F_A(q^2 = 0; n \to p) \sum_{k=1}^{3} \tau_k^{(+)} \boldsymbol{\sigma}_k | \Psi_{M_i}(H^3) \rangle \right|^2 \right\}^{1/2},$$

$$F_A(q^2 = 0; n \to p) = 1.23 \pm 0.01 \tag{3}$$

where $\Psi_{M_i}(H^3)$ and $\Psi_{M_f}(He^3)$ are wavefunctions for H^3 and He^3 depending on the nucleon position, spin, and isospin coordinates $(\ldots \boldsymbol{r}_k, \sigma_k^{(3)}, \tau_k^{(3)} \ldots)$, $J = 1/2$ is the spin of He^3 (and also of H^3), and $\sum_{k=1}^{3} \tau_k^{(+)} \boldsymbol{\sigma}_k$ is the model's axial hadronic weak current operator. With the "best available" $\Psi_{M_i}(H^3)$ and $\Psi_{M_f}(He^3)$ one finds [2]

$$\sum_{M_f = -J}^{J} \left| \langle \Psi_{M_f}(He^3) | \sum_{k=1}^{3} \tau_k^{(+)} \boldsymbol{\sigma}_k | \Psi_{M_i}(H^3) \rangle \right|^2 = 2.61 \pm 0.10\,, \tag{4}$$

so that

$$|F_A(q^2 = 0; H^3 \to He^3)| = F_A(q^2 = 0; n \to p)\,(0.93 \pm 0.02) = 1.14 \pm 0.02 \tag{5}$$

which is 6% smaller than the corresponding measured value of Eq. (2). This 6% discrepancy is generally believed to demonstrate the quantitative inadequacy of the conventional neutron-proton nuclear model in semileptonic weak processes since, as indicated by the assigned error in Eq. (4), the uncertainty in the "best available" $\Psi_{M_i}(H^3)$ and $\Psi_{M_f}(He^3)$ corresponds to no more than a 2% uncertainty in Eq. (5). Similar conclusions regarding the quantitative inadequacy of the conventional neutron-proton nuclear model in electromagnetic processes may be drawn on the basis of a comparison of measured and model-predicted H^3 and He^3 magnetic moments; one has [3]

$$[\mu(H^3)-\mu(He^3)]_{\text{meas.}}=5.11\left(\frac{e}{2m_p}\right) \tag{6}$$

versus

$$[\mu(H^3)-\mu(He^3)]_{n-p\ \text{nuc. mod.}}=(4.37\pm 0.08)\left(\frac{e}{2m_p}\right) \tag{7}$$

which is a 14% discrepancy. On the other hand, no reliable estimate of the quantitative inadequacy of the conventional neutron-proton nuclear model in weak semileptonic or electromagnetic processes can be provided at present either for deuterium or for mirror nuclei heavier than $[H^3, He^3]$ – this sad circumstance is a consequence of the fact that for the relatively loosely bound deuteron the inadequacy is so small that various hard-to-estimate relativistic effects must also be considered while for the relatively complex mirror nuclei heavier than $[H^3, He^3]$ the model's wavefunctions, $\Psi_{M_i}(N_i)$ and $\Psi_{M_f}(N_f)$, are not well enough known to permit a sufficiently precise calculation of

$$\sum_{M_f=-J}^{J}\left|\langle\Psi_{M_f}(N_f)|\sum_{k=1}^{A}\tau_k^{(+)}\boldsymbol{\sigma}_k|\Psi_{M_i}(N_i)\rangle\right|^2 \quad\text{or}\quad [\mu(N_i)-\mu(N_f)].$$

The explanation of the discrepancy between $|F_A(q^2=0;H^3\to He^3)|_{\text{meas.}}$ (Eq. (2)) and $|F_A(q^2=0;H^3\to He^3)|_{n-p\ \text{nuc. mod.}}$ (Eq. (5)) (or between $[\mu(H^3)-\mu(He^3)]_{\text{meas.}}$ (Eq. (6)) and $[\mu(H^3)-\mu(He^3)]_{n-p\ \text{nuc. mod.}}$ (Eq. (7))) can be given in two different ways; a treatment of these explanations deeper than so far attempted is required to understand fully the relationship between them. In the first of these explanations one notes that the expression for $|F_A(q^2=0;H^3\to He^3)|_{n-p\ \text{nuc. mod.}}$ in Eq. (3) contains contributions from processes such as $n_k\to p_k+e^-+\nu_e$ only; therefore, augmenting the contributions from these processes by contributions from processes such as $n_k\to n_k+\pi^++e^-+\nu_e$; $\pi^++n_l\to p_l$ etc., one may hope to remove the discrepancy. In a first and crude approximation it is clear that inclusion of such "meson-exchange" corrections results in the replacement of $\sum_{k=1}^{3} \tau_k^{(+)}\boldsymbol{\sigma}_k$ by

$$\sum_{k=1}^{3} \tau_k^{(+)}\boldsymbol{\sigma}_k + \frac{f^2}{4\pi} \sum_{k=1,l=1}^{3} (a(\boldsymbol{\tau}_k \times \boldsymbol{\tau}_l)^{(+)} (\boldsymbol{\sigma}_k \times \boldsymbol{\sigma}_l) + b(\tau_k^{(+)}\boldsymbol{\sigma}_k + \tau_l^{(+)}\boldsymbol{\sigma}_l) + c(\tau_k^{(+)}\boldsymbol{\sigma}_l + \tau_l^{(+)}\boldsymbol{\sigma}_k)) \frac{e^{-m_\pi r_{kl}}}{m_\pi r_{kl}} \tag{8}$$

where $f = \sqrt{4\pi(0.08)}$ is the coupling constant associated with the $p \to p + \pi^0$ process and $|a|$, $|b|$, $|c|$ are quantities ≈ 1 at $r_{kl} \cong (m_\pi)^{-1}$ – this replacement essentially removes the discrepancy under consideration. A rather elaborate recent calculation, which employs PCAC [4] to evaluate the axial form factors associated with $n_k \to n_k + \pi^+ + e^- + \nu_e$, etc., confirms this removal [5].

In the second of the explanations addressed to the discrepancy between the measured and the model-predicted values of

$$|F_A(q^2 = 0; \mathrm{H}^3 \to \mathrm{He}^3)|$$

one assumes that the various nuclear wavefunctions contain relatively small terms describing the presence of bound nucleon-isobars in addition to the main terms describing the presence of bound nucleons. Thus one contemplates contributions to the predicted $|F_A(q^2 = 0; \mathrm{H}^3 \to \mathrm{He}^3)|$ arising from processes such as $N_k^{*0} \to p_k + e^- + \nu_e$ and $n_k \to N_k^{*+} + e^- + \nu_e$ where, for illustrative purposes, we have supposed that the nucleon-isobar in question has the same spin and isospin as the nucleon (i.e. N^*(1470 Mev)). To work out these contributions one introduces the replacements

$$\begin{aligned} \Psi_{M_i}(\mathrm{H}^3) &\to \sqrt{1-|C|^2}\, \Psi_{M_i}(\mathrm{H}^3) + C\Phi_{M_i}(\mathrm{H}^3) \\ \Psi_{M_f}(\mathrm{He}^3) &\to \sqrt{1-|C|^2}\, \Psi_{M_f}(\mathrm{He}^3) + C\Phi_{M_f}(\mathrm{He}^3) \\ C &\equiv \frac{\langle \Phi_{M_i}(\mathrm{H}^3)| \sum_{k=1,l=1}^{3} V(N_k, N_l^*) \,|\, \Psi_{M_i}(\mathrm{H}^3)\rangle}{(m_N - m_{N^*})} \end{aligned} \tag{9}$$

and

$$\sum_{k=1}^{3} \tau_k^{(+)}\boldsymbol{\sigma}_k \to \sum_{k=1}^{3} t_k^{(+)}\boldsymbol{\sigma}_k \tag{10}$$

where $\Phi_{M_i}(\mathrm{H}^3)$ and $\Phi_{M_f}(\mathrm{He}^3)$ are wavefunctions depending on the position, spin and isospin coordinates of one nucleon-isobar and two nucleons, $V(N_k, N_l^*)$ is the interaction potential energy associated with the process $N_k + N_l \to N_k + N_l^*$ ($N_k \equiv p_k$ or n_k) and is presumably such that the probability of finding the N^* in the nucleus, $|C|^2$, is small (see

Eqs. (12–14) below), and the operator $t_k^{(+)}$ is a suitable generalization of the operator $\tau_k^{(+)}$ [6]. This replacement leads to a modification of Eqs. (3)–(5) as follows:

$$|F_A(q^2=0;\mathrm{H}^3\to\mathrm{He}^3)|\cong(1.14\pm0.02)\,|(1+CX)|$$

$$X\equiv\frac{2\sum\limits_{M_f=-J}^{J}\left(\mathrm{Re}\langle\Psi_{M_f}(\mathrm{He}^3)|\sum\limits_{k=1}^{3}\tau_k^{(+)}\sigma_k|\Psi_{M_i}(\mathrm{H}^3)\rangle\right)\cdot\left(\mathrm{Re}\langle\Psi_{M_f}(\mathrm{He}^3)|\sum\limits_{k=1}^{3}t_k^{(+)}\sigma_k|\Phi_{M_i}(\mathrm{H}^3)\rangle\right)}{\sum\limits_{M_f=-J}^{J}\left|\langle\Psi_{M_f}(\mathrm{He}^3)|\sum\limits_{k=1}^{3}\tau_k^{(+)}\sigma_k|\Psi_{M_i}(\mathrm{H}^3)\rangle\right|^2} \tag{11}$$

whence

$$X\approx2\frac{F_A(q^2=0;N^{*0}\to p)}{F_A(q^2=0;n\to p)}\cong2\frac{f^*}{f} \tag{12}$$

where the second (approximate) equality is a consequence of the Goldberger-Treiman relation, $f^*=\sqrt{4\pi(0.01)}$ being the coupling constant associated with the $N^{*+}\to p+\pi^0$ process [7]. Further

$$C\approx\frac{f^*}{f}\,\frac{\langle\Psi_{M_i}(\mathrm{H}^3)|\sum\limits_{k=1,l=1}^{3}V(N_k,N_l)|\Psi_{M_i}(\mathrm{H}^3)\rangle}{(m_N-m_{N^*})}\cong\frac{f^*}{f}\left(\frac{80\,\mathrm{Mev}}{530\,\mathrm{Mev}}\right) \tag{13}$$

where $2V(N_k,N_l)$ is the interaction potential energy associated with the process $N_k+N_l\to N_k+N_l$. Thus, combining Eqs. (12) and (13),

$$CX\approx0.3\left(\frac{f^*}{f}\right)^2=0.04 \tag{14}$$

which is of the same order of magnitude as the value $CX=0.06$ required for the agreement between the predicted value of $|F_A(q^2=0;\mathrm{H}^3\to\mathrm{He}^3)|$ in Eq. (11) and the measured value of $|F_A(q^2=0;\mathrm{H}^3\to\mathrm{He}^3)|$ in Eq. (2). It thus appears as if the conventional neutron-proton nuclear model, augmented by an appropriate treatment of "bound nucleon-isobar" effects, may eventually offer a quantitatively precise description of semi-leptonic weak and electromagnetic processes in nuclei – clearly, much additional theoretical work will have to be done along the lines sketched in Eqs. (9)–(14) before this conclusion can be considered as firmly established [8].

Once we adjust our thinking to the notion that nuclei contain the various nucleon-isobars as well as nucleons we can contemplate processes which are impossible from the point of view of the conventional neutron-proton nuclear model, e.g. the processes of muon capture from an atomic orbit:

$$\mu^- + \begin{pmatrix} p_{\text{bound}} \\ n_{\text{bound}} \end{pmatrix} \rightarrow \nu_\mu + \begin{pmatrix} N^{*0}_{\text{bound}} \\ N^{*-}_{\text{bound}} \end{pmatrix} \tag{15}$$

and

$$\mu^- + \begin{pmatrix} N^{*++}_{\text{bound}} \\ N^{*+}_{\text{bound}} \end{pmatrix} \rightarrow \nu_\mu + \begin{pmatrix} p_{\text{bound}} \\ n_{\text{bound}} \end{pmatrix} \tag{16}$$

and the processes induced by an incident neutrino:

$$\nu_\mu + \begin{pmatrix} p_{\text{bound}} \\ n_{\text{bound}} \end{pmatrix} \rightarrow \mu^- + \begin{pmatrix} N^{*++}_{\text{bound}} \\ N^{*+}_{\text{bound}} \end{pmatrix} \tag{17}$$

and

$$\nu_\mu + \begin{pmatrix} N^{*0}_{\text{bound}} \\ N^{*-}_{\text{bound}} \end{pmatrix} \rightarrow \mu^- + \begin{pmatrix} p_{\text{bound}} \\ n_{\text{bound}} \end{pmatrix} \tag{18}$$

where, in all cases, the nucleon and the nucleon-isobar are bound in their respective nuclei; as indicated by our assignment of charge states to the nucleon-isobar, we suppose here that this nucleon isobar has $I=3/2$ and in fact we take $N^* \equiv N^*$ (1236 Mev; $J=3/2, I=3/2$). The processes of Eqs. (15)–(18) are characterized by some interesting features not shared by the normally dominant muon-capture and neutrino-induced processes

$$\mu^- + p_{\text{bound}} \rightarrow \nu_\mu + n_{\text{bound}}\,, \tag{19}$$

$$\nu_\mu + n_{\text{bound}} \rightarrow \mu^- + p_{\text{bound}}\,, \tag{20}$$

where the nucleons involved are again bound in nuclei. Thus, if the $\Delta S=0$ hadronic weak current possesses an isotensor component $h_\lambda^{(\Delta I=2)}$, in addition to the standard isovector component $h_\lambda^{(\Delta I=1)}$, the isotensor component will contribute to the nucleon → nucleon-isobar and nucleon-isobar → nucleon transitions but not to the nucleon → nucleon transitions. Therefore, while awaiting the results of detailed measurements on $\nu_\mu + p \rightarrow \mu^- + N^{*++}$ in hydrogen, one could do worse than study various suitable $\mu^- + N_{\text{ini}} \rightarrow \nu_\mu + N_{\text{fin}}$ and $\nu_\mu + N_{\text{ini}} \rightarrow \mu^- + N_{\text{fin}}$ reactions with an eye toward isolation of contributions from the processes of Eqs. (15) and (16) and Eqs. (17) and (18). In particular, observation of a reaction such as

$$\begin{gathered} \mu^- + \mathrm{He}^4(J=0^{(+)}, I=0) \rightarrow \nu_\mu + \mathrm{H}^4(J=0^{(+)}, I=2) \\ \mathrm{H}^4(J=0^{(+)}, I=2) \rightarrow n+n+n+p\,, \end{gathered} \tag{21}$$

with a rate significantly greater than that calculated from the matrix element $\langle \mathrm{H}^4(J=0^{(+)}, I=2)\,|h_\lambda^{(\Delta I=1)}|\, I=1$ and $I=2$ impurity in $\mathrm{He}^4(J=0^{(+)}, I=0)\rangle$, would be sufficient to establish the existence of $h_\lambda^{(\Delta I=2)}$ [9].

II b) Semileptonic Weak Processes in Nuclei — Nuclear Two-Neutrino and No-Neutrino Double-Beta Decay

In the previous section we discussed some semileptonic weak processes in nuclei which, on the one hand, demonstrated the quantitative inadequacy of the conventional neutron-proton nuclear model, and, on the other hand, could conceivably yield information about the hadronic weak current which is as yet unavailable. In this section we shall focus our attention on the leptonic weak current and discuss whether this current is of such a form as to insure the rigorous validity of the principle of lepton conservation. Our discussion will, of course, be guided by whatever experimental evidence exists in favor of this principle; the evidence in question involves the non-observation of the neutrino-induced reactions [10]

$$\nu_e + n_{\text{bound}} \to e^- + p_{\text{bound}}, \tag{22}$$

$$\nu_\mu + p_{\text{bound}} \to \mu^+ + n_{\text{bound}}, \tag{23}$$

where the ν_e and ν_μ originate from $n_{\text{bound}} \to p_{\text{bound}} + e^- + \nu_e$ and $\pi^+ \to \mu^+ + \nu_\mu$, respectively, and the theoretical analysis of the measured rates of nuclear double-beta decays from which conclusions can be drawn regarding whether two neutrinos do or do not accompany the two electrons. In particular, we shall concentrate our attention on the nuclear double-beta decay reactions and begin by emphasizing that, recently, several of these reactions have been definitely observed by *Kirsten* and his collaborators [11], viz.:

$$\text{Te}^{130} \to \text{Xe}^{130} + e^- + e^- + (\nu_e + \nu_e), \tag{24}$$

$$\text{Se}^{82} \to \text{Kr}^{82} + e^- + e^- + (\nu_e + \nu_e). \tag{25}$$

There is also some not entirely unambiguous evidence for the nuclear double-beta decay reaction [12]

$$\text{Te}^{128} \to \text{Xe}^{128} + e^- + e^- + (\nu_e + \nu_e). \tag{26}$$

Then, if the two neutrinos always accompany the two electrons, lepton conservation is valid, while if the two electrons *sometimes* appear without the two neutrinos, lepton conservation does not hold. The observation of the nuclear double-beta decay processes in Eqs. (24)–(26) has up to now been indirect – the daughter atoms have been detected by a mass-spectrometric technique in rocks of known age containing the parent atoms, and mechanisms for the production of the daughter atoms alternative to nuclear double-beta decay (e.g. by spontaneous fission of uranium atoms present as impurities in the rocks) excluded by rather convincing arguments (except possibly in the case of Eq. (26) – see below). However in this method the two electrons are not observed, and one cannot make the distinction between no-neutrino and two-neutrino double-beta decay on the basis of whether or not the sum of the energies of the two electrons is the same in all decays. The no-neutrino vs. two-neutrino distinction is nevertheless still possible since (1) the *ratio* of the double-beta decay rates of two isotopes such as Te^{130} and Te^{128} should be approximately equal to the *ratio* of the corresponding phase-spaces and since (2) the four-lepton and two-lepton phase-spaces vary approximately as $E^{8.4}$ and $E^{4.2}$, respectively, where E is the energy release in the decay (the phase-space for a state with n zero-mass leptons varies as E^{3n-1}). The energy releases are 2.5 Mev for Te^{130} and 0.85 Mev for Te^{128}, a ratio of about 3 to 1, and so the predicted values for the ratios of the life-times are

$$\frac{T(\mathrm{Te}^{128})}{T(\mathrm{Te}^{130})} \cong 3^{8.4} = 10^4 : \text{four-lepton final state,} \tag{27}$$

$$\frac{T(\mathrm{Te}^{128})}{T(\mathrm{Te}^{130})} \cong 3^{4.2} = 10^2 : \text{two-lepton final state.} \tag{28}$$

Experimentally,

$$T(\mathrm{Te}^{128}) \cong 10^{22.5} \text{ years}, \tag{29}$$

$$T(\mathrm{Te}^{130}) = 10^{21.3} \text{ years}, \tag{30}$$

which obviously favors the ratio appropriate to the two-lepton final state and hence constitutes evidence for the no-neutrino process, and so evidence for lepton nonconservation. The theoretical argument here is fairly reliable as it is based on the idea that the $Te^{130} \to Xe^{130}$ and $Te^{128} \to Xe^{128}$ nuclear matrix elements have about the same magnitude. Thus, the validity of the conclusion that lepton conservation is violated rests *almost entirely* on the assumption that the observed Xe^{128} is produced predominantly by the double-beta decay of Te^{128} and not, e.g., by the (α, n) nuclear reaction: $Te^{125} + \alpha \to Xe^{128} + n$ with α-particles emitted by naturally radioactive impurity atoms in the Te rock [13].

Assuming for the sake of the further discussion that the nuclear double-beta decays in Eqs. (24)–(26) are indeed predominantly no-neutrino, the corresponding lepton nonconservation parameter can be calculated. The mechanism of the no-neutrino nuclear double-beta decay process may be schematically represented as, e.g..

$$\mathrm{Te}^{130} \to (\mathrm{I}^{130})_n + e^- + \nu_e \to \mathrm{Xe}^{130} + e^- + e^- \tag{31}$$

where $(\mathrm{I}^{130})_0, (\mathrm{I}^{130})_1, \ldots, (\mathrm{I}^{130})_n, \ldots$ are the ground and the various excited states of the intermediate nucleus, and where, to obtain a nonvanishing expression for the $Te^{130} \to Xe^{130} + e^- + e^-$ matrix element, the leptonic weak current $l_\lambda^{(e)}(\boldsymbol{x})$ must be taken as [14]

$$l_\lambda^{(e)}(\boldsymbol{x}) = (\overline{\psi}_e(\boldsymbol{x})\, \gamma_\lambda \{(1+\gamma_5) + \eta(1-\gamma_5)\}\, \psi_{\nu_e}(\boldsymbol{x}))\,, \qquad \tilde{\psi}_{\nu_e}(\boldsymbol{x})^\dagger = \psi_{\nu_e}(\boldsymbol{x})\,. \tag{32}$$

The constant η which may be complex (*CP* or *T* noninvariance!!), but which we here suppose to be real, determines the relative weight of the two helicity states accessible to the neutrino and thus measures the extent of lepton nonconservation. We note that the condition: $\tilde{\psi}_{\nu_e}^\dagger = \psi_{\nu_e}$ implies that the neutrino is taken to be a "Majorana" particle; if the neutrino were a "Dirac" particle: $\psi_{\bar{\nu}_e} \equiv \tilde{\psi}_{\nu_e}^\dagger \neq \psi_{\nu_e}$, lepton conservation would hold even with $\eta \neq 0$.

The weak-interaction Hamiltonian for semileptonic strangeness-preserving processes

$$H_{\text{weak s.l.}}^{(\Delta S=0)} = \frac{G}{\sqrt{2}} \int l_\lambda^{(e)}(\boldsymbol{x})\;\; h_\lambda^\dagger(\boldsymbol{x})\, d\boldsymbol{x} + \text{herm conj}\,,$$
$$h_\lambda(\boldsymbol{x}) = V_\lambda(\boldsymbol{x}) + A_\lambda(\boldsymbol{x})\,, \tag{33}$$
$$G = G_\mu \cos\vartheta = \left(\frac{1.02 \times 10^{-5}}{m_p^2}\right)(0.98) = \frac{1.00 \times 10^{-5}}{m_p^2}$$

with $V_\lambda(x)$ and $A_\lambda(x)$ polar and axial $\Delta S = 0$ hadronic weak currents, and with $l_\lambda^{(e)}(\boldsymbol{x})$ given by Eq. (32), yields an effective coupling proportional to G at one lepton-nucleus-nucleus vertex in Eq. (31) and proportional to $G\eta$ at the other such vertex. The resultant no-neutrino nuclear double-beta decay amplitude is then proportional to $G^2\eta$ and the corresponding decay rate is given by

$$\Gamma(Te^{130} \to Xe^{130} + e^- + e^-) \approx G^2 E^5 \left(\frac{G\eta \langle E_\nu \rangle^3}{\Delta E + \langle E_\nu \rangle}\right)^2 \tag{34}$$

where the $G^2 E^5$ corresponds to a single-beta decay at one lepton-nucleus-nucleus-vertex, ΔE is an average energy arising from the summation over the intermediate I^{130} states, and $\langle E_\nu \rangle$ is the average energy of the virtual neutrino. In nuclear double-beta decay both electrons are negative and so *cannot* be emitted by the *same* nucleon (or more generally, by the *same* basic nuclear constituent as long as all such basic constituents are assumed to posses only two charge states); thus

$$\langle E_\nu \rangle \approx \frac{1}{\langle r \rangle} \approx \frac{m_\pi}{A^{1/3}} \gg \Delta E \tag{35}$$

where $\langle r \rangle$ is the average spacing between the nucleon which emits the neutrino and the first electron and the nucleon which reabsorbs the neutrino and emits the second electron. (When the charged leptons emitted have opposite charges as, e.g., in $\pi^0 \rightarrow e^- + e^+$ or $K_L^0 \rightarrow \mu^- + \mu^+$, the *same* basic meson constituent (quark) *can* emit the neutrino and the first charged lepton and then reabsorb the neutrino and emit the second charged lepton; $\langle r \rangle$ is then effectively zero and the second-order weak amplitudes for $\pi^0 \rightarrow e^- + e^+$ and $K_L^0 \rightarrow \mu^- + \mu^+$ diverge. In this case the divergence is avoided only if one introduces a damping or cancellation mechanism extraneous to the $H_{\text{weak s.l.}}^{(\Delta S=0)}$ of Eq. (33).)

For the nuclear double-beta decay with two neutrinos emitted the virtual neutrino loop is absent and so there is no possibility of a divergence. The decay rate is given by

$$\Gamma(\mathrm{Te}^{130} \rightarrow \mathrm{Xe}^{130} + e^- + e^- + \nu_e + \nu_e) \approx G^2 E^5 \left(\frac{GE^3}{\Delta E}\right)^2 \tag{36}$$

and this yields a lifetime $\approx 10^{21}$ years for $E = 2.5$ Mev, in agreement with the experimental value in Eq. (30). In fact, if only the Te^{130} and Se^{82} data were available, the two-neutrino double-beta decay theory, which corresponds to lepton conservation, would appear satisfactory; however, as emphasized above, this theory predicts an incorrect value for the ratio of the lifetimes of Te^{128} and Te^{130}. The no-neutrino double-beta decay theory, which corresponds to lepton nonconservation, predicts this ratio more or less correctly, and the lepton nonconservation parameter η is fixed at

$$\eta \approx 10^{-3} \tag{37}$$

by equating the calculated Te^{130} double-beta decay rate (Eqs. (34) and (35)) with the reciprocal of the observed Te^{130} lifetime (Eq. (30)). This value of η is much smaller than the limit that can be set on it from any of the single-beta processes, e.g. much smaller than the limit $\eta < 0.2$ deduced from the non-observation of $\nu_e + n_{\text{bound}} \rightarrow e^- + p_{\text{bound}}$ with ν_e from decays of the type $n_{\text{bound}} \rightarrow p_{\text{bound}} + e^- + \nu_e$ [10]. It is also worth mentioning that with the $l_\lambda^{(e)}(x)$ of Eq. (32), the $H_{\text{weak s.l.}}^{(\Delta S=0)}$ of Eq. (33) is no longer invariant under the transformation $\psi_{\nu_e} \rightarrow \gamma_5 \psi_{\nu_e}$ so that the neutrino will acquire a non-vanishing weak-interaction self-energy and a corresponding mass m_{ν_e}. One has

$$\frac{m_{\nu_e}}{m_e} \approx G^2 \eta M^4 \tag{38}$$

where M is a cut-off mass; the experimental limit: $m_{\nu_e}/m_e < 10^{-4}$ implies that $M < 100$ Gev.

II c) Semileptonic Weak Processes in Nuclei — "Second-Class" Hadronic Weak Currents and Time-Reversal Noninvariance

At present there is no definite experimental evidence for CP or T noninvariance other than that found in the pionic and leptonic K_L^0 decays. In a world where the CP noninvariance arises from $H_{\text{superweak}}$ one would in fact expect to find CP noninvariance effects only in the K_L^0, K_S^0 system. On the other hand, many crucial experiments on CP or T noninvariance have either not yet been performed or have been performed with insufficient accuracy so that it may turn out that CP or T noninvariance characterizes weak processes other than the K_L^0 decays, e.g., the weak semileptonic strangeness-preserving processes. If this happens, one will be forced to assume (see Eq. (33)) that either the $l_\lambda^{(e)}(x)$ or the $h_\lambda(x) = (V_\lambda(x) + A_\lambda(x))$ or both possess a "CP-abnormal" piece; in what follows we shall consider some of the consequences of such an assumption.

As we have already indicated, the lepton-nonconservation parameter η which enters into the $l_\lambda^{(e)}(x)$ of Eq. (32) can be assumed complex so that [15]

$$l_\lambda^{(e)}(x) = \{l_\lambda^{(e)}(x)\}_{CP\text{-normal}} + \{l_\lambda^{(e)}(x)\}_{CP\text{-normal}} \tag{39}$$

with

$$\begin{aligned}\{l_\lambda^{(e)}(x)\}_{CP\text{-normal}} &= (\bar{\psi}_e(x)\,\gamma_\lambda\{(1+\gamma_5) + \mathrm{Re}\,\eta(1-\gamma_5)\}\,\psi_{\nu_e}(x)),\\ \{l_\lambda^{(e)}(x)\}_{CP\text{-abnormal}} &= i\,\mathrm{Im}\,\eta(\bar{\psi}_e(x)\,\gamma_\lambda(1-\gamma_5)\,\psi_{\nu_e}(x)).\end{aligned} \tag{40}$$

Then, Eq. (37) reads $|\eta| \approx 10^{-3}$, and if one further assumes that

$$\mathrm{Im}\,\eta \cong \mathrm{Re}\,\eta \tag{41}$$

one establishes a quantitative relation between the extent of lepton nonconservation and the extent of T noninvariance. Thus, with this assumption, one may anticipate various T-noninvariant effects in nuclear single-beta decay, e.g., the appearance of a transverse polarization for the emitted electrons ($\sim \mathrm{Im}\,\eta[\boldsymbol{S}_e \cdot (\boldsymbol{p}_e \times \boldsymbol{p}_{ev})]$) on the level of 0.1%. Of course, the presence of T-noninvariant effects of this order would be extremely difficult to establish and would probably require quite novel experimental techniques.

A less forbidding prospect from the point of view of an eventual experimental verification arises if one assumes [16]

$$\begin{aligned}V_\lambda(x) &= \{V_\lambda(x)\}_{CP\text{-normal}} = I^{(-)}(x),\\ A_\lambda(x) &= \{A_\lambda(x)\}_{CP\text{-normal}} + \{A_\lambda(x)\}_{CP\text{-abnormal}},\end{aligned} \tag{42a}$$

where, as required by CVC, $I_\lambda^{(-)}(x)$ is the indicated component of the iso-current, and where, without any essential loss in the generality of our argument, we can also assume that both $\{A_\lambda(x)\}_{CP\text{-normal}}$ and $\{A_\lambda(x)\}_{CP\text{-abnormal}}$ transform under rotations in isospace like $I_\lambda^{(-)}(x)$ and are therefore first-class currents and second-class currents respectively. Finally, we assume that in some well-defined sense

$$\{A_\lambda(x)\}_{CP\text{-abnormal}} \approx \{A_\lambda(x)\}_{CP\text{-normal}} \tag{42b}$$

so that the T noninvariance is maximal.

Consider now the beta decays

$$\begin{aligned} n &\rightarrow p \quad + e^- + \nu_e\,, \\ \mathrm{Ne}^{19} &\rightarrow \mathrm{F}^{19} + e^+ + \nu_e\,. \end{aligned} \tag{43}$$

In this case, careful experiments have been performed to find the T-non-invariant asymmetry:

$$\left(\frac{2F_V(\alpha\rightarrow\beta)\,\mathrm{Im}\,F_A(\alpha\rightarrow\beta)}{(F_V(\alpha\rightarrow\beta))^2+3|F_A(\alpha\rightarrow\beta)|^2}\right)[\langle S_\alpha\rangle\cdot(p_e\times p_\nu)] \tag{44}$$

where $F_V(\alpha\rightarrow\beta)$ and $F_A(\alpha\rightarrow\beta)=\mathrm{Re}\,F_A(\alpha\rightarrow\beta)+i\,\mathrm{Im}\,F_A(\alpha\rightarrow\beta)$ are the polar and axial form factors at zero momentum transfer given by

$$\begin{aligned} F_V(\alpha\rightarrow\beta) &= \frac{\langle\beta|I_0^{(+)}(0)|\alpha\rangle}{(u_\beta^\dagger\,\tau^{(+)}u_\alpha)}\,, \\ \mathrm{Re}\,F_A(\alpha\rightarrow\beta) &= \frac{\langle\beta|\{A_3(0)\}^\dagger_{CP\text{-normal}}|\alpha\rangle}{(u_\beta^\dagger\,\tau^{(+)}\sigma_3\,u_\alpha)}\,, \\ i\,\mathrm{Im}\,F_A(\alpha\rightarrow\beta) &= \frac{\langle\beta|\{A_3(0)\}^\dagger_{CP\text{-abnormal}}|\alpha\rangle}{(u_\beta^\dagger\,\tau^{(+)}\sigma_3\,u_\alpha)} \end{aligned} \tag{45}$$

with $\alpha=n$, $\beta=p$, or $\alpha=\mathrm{Ne}^{19}$, $\beta=\mathrm{F}^{19}$ [17]. Experimentally, there is no asymmetry observed either in the case $n\rightarrow p$ or in the case $\mathrm{Ne}^{19}\rightarrow\mathrm{F}^{19}$, and one can set the limit

$$\frac{|\mathrm{Im}\,F_A(\alpha\rightarrow\beta)|}{|F_A(\alpha\rightarrow\beta)|}<0.01\,. \tag{46}$$

Offhand, this would seem to demonstrate that

$$\{A_\lambda(x)\}_{CP\text{-abnormal}} \ll \{A_\lambda(x)\}_{CP\text{-normal}}$$

and so contradict Eq. (42b). However, one can show (see below), that if the states $|\alpha\rangle$ and $|\beta\rangle$ are the two members of an isodoublet, as are $|n\rangle$ and $|p\rangle$ and also $|\mathrm{Ne}^{19}\rangle$ and $|\mathrm{F}^{19}\rangle$, then

$$\langle\beta|\,\{A_3(0)\}^\dagger_{CP\text{-abnormal}}\,|\alpha\rangle = \langle\beta|\,\{A_3(0)\}^\dagger_{CP\text{-abnormal}}\,|\alpha\rangle^*\,. \tag{47}$$

Combination of Eqs. (47) and (45), together with the reality of $(u_\beta^\dagger \tau^{(+)} \sigma_3 u_\alpha)$, yields

$$\operatorname{Im} F_A(\alpha \to \beta) = 0 \tag{48}$$

for the processes of Eq. (43) even if $\operatorname{Im} F_A(\alpha \to \beta) \neq 0$ for nuclear beta decays in general. Hence, for an *unambiguous* test of T noninvariance in nuclear beta decay, one must search for an asymmetry such as that in Eq. (44) in transitions between two nuclear states which are *not* both members of the same isodoublet: a good candidate is [16]

$$\mathrm{P}^{32} \to \mathrm{S}^{32} + e^- + \nu_e \tag{49}$$

and it is to be hoped that T-noninvariance experiments with this or with other similar nuclear beta decays will be undertaken soon.

The reason that the nuclear beta decay in Eq. (49) is promising from the point of view of an unambiguous test of Eq. (42b) is the following: On the basis of the conventional neutron-proton nuclear model as corrected by meson-exchange terms one has

$$\frac{|\operatorname{Im} F_A(\alpha \to \beta)|}{|F_A(\alpha \to \beta)|} = \frac{|\operatorname{Im} F_A(n \to p) \langle \tau^{(+)} \sigma \rangle_{\alpha \to \beta} + \operatorname{Im} f_A(\alpha \to \beta)|}{|F_A(n \to p) \langle \tau^{(+)} \sigma \rangle_{\alpha \to \beta} + f_A(\alpha \to \beta)|} \tag{50}$$

where

$$\langle \tau^{(+)} \sigma \rangle_{\alpha \to \beta} \equiv \pm \left\{ \left(\frac{J_\alpha}{J_\alpha + 1} \right) \sum_{M_\beta = -J_\beta}^{J_\beta} \left| \langle \Psi_{M_\beta}(\beta) | \sum_{k=1}^{A} \tau_k^{(+)} \boldsymbol{\sigma}_k | \Psi_{M_\alpha}(\alpha) \rangle \right|^2 \right\}^{1/2} \tag{51}$$

and $f_A(\alpha \to \beta)$ is the meson-exchange correction to $F_A(\alpha \to \beta)$; according to Section II a) above one expects $|\operatorname{Re} f_A(\alpha \to \beta)|$ to be of order $10^{-1} - 10^{-2}$. Thus, remembering that $\operatorname{Im} F_A(n \to p) = 0$ (Eq. (48)), Eq. (50) becomes

$$\frac{|\operatorname{Im} F_A(\alpha \to \beta)|}{|F_A(\alpha \to \beta)|} = \frac{|\operatorname{Im} f_A(\alpha \to \beta)|}{[F_A^2(n \to p) \langle \tau^{(+)} \sigma \rangle_{\alpha \to \beta}^2 + 2 F_A(n \to p) \langle \tau^{(+)} \sigma \rangle_{\alpha \to \beta} \operatorname{Re} f_A(\alpha \to \beta) + |f_A(\alpha \to \beta)|^2]^{1/2}} \tag{52}$$

and we see that $\dfrac{|\operatorname{Im} F_A(\alpha \to \beta)|}{|F_A(\alpha \to \beta)|}$ can be of order unity only if

$$|\operatorname{Im} f_A(\alpha \to \beta)| \approx |\operatorname{Re} f_A(\alpha \to \beta)| \tag{53}$$

and

$$F_A(n \to p) |\langle \tau^{(+)} \sigma \rangle_{\alpha \to \beta}| = 1.2 |\langle \tau^{(+)} \sigma \rangle_{\alpha \to \beta}| \lesssim |\operatorname{Re} f_A(\alpha \to \beta)| \,. \tag{54}$$

Now Eq. (53) may be expected to hold essentially for all $\alpha \to \beta + e + \nu_e$ single-beta decays if Eq. (42b) is valid; on the other hand, in view of $|\operatorname{Re} f_A(\alpha \to \beta)|$ being of order $10^{-1} - 10^{-2}$, Eq. (54) holds only if

$|F_A(\alpha \to \beta)| < 10^{-1} - 10^{-2}$ and this condition is known from experiment to be valid for $P^{32} \to S^{32} + e^- + \nu_e$ ($|F_A(P^{32} \to S^{32})| = 0.6 \times 10^{-2}$) and for several similar "*l*-forbidden" nuclear beta decays.

We conclude this section with the above-promised proof of Eq. (47). We have, since $\{A_\lambda(x)\}_{CP\text{-abnormal}} \equiv a_\lambda(x)$ transforms under rotations in isospace like $I_\lambda^{(-)}(x)$,

$$R a_3^\dagger(x) R^{-1} = -a_3(x),$$

$$R \equiv \exp(-i\pi I^{(2)}) = \exp\left(-\frac{\pi}{2} \int [I_0^{(+)}(x) - I_0^{(-)}(x)]\, \mathrm{d}x\right) \tag{55}$$

and from this, when in addition $|\alpha\rangle$ and $|\beta\rangle$ are the two members of the same isodoublet, i.e. when

$$|\alpha_R\rangle \equiv R|\alpha\rangle = -|\beta\rangle, \qquad |\beta_R\rangle \equiv R|\beta\rangle = |\alpha\rangle \tag{56}$$

we have

$$\begin{aligned} \langle\beta|a_3^\dagger(0)|\alpha\rangle &= \langle\beta_R|R a_3^\dagger(0) R^{-1}|\alpha_R\rangle \\ &= \langle\alpha_R|(R a_3^\dagger(0) R^{-1})^\dagger|\beta_R\rangle^* \\ &= \langle\alpha_R|-a_3^\dagger(0)|\beta_R\rangle^* \qquad \text{(using Eq. (55))} \\ &= \langle-\beta|-a_3^\dagger(0)|\alpha\rangle^* \qquad \text{(using Eq. (56))} \end{aligned} \tag{57}$$

which is just Eq. (47).

II d) Semileptonic Weak Processes in Nuclei — Goldberger-Treiman Relation, Adler-Weisberger Sum Rule, Weinberg-Tomozawa Formula

In order to broaden the area of application of PCAC, and ETCR (Equal Time Commutation Relations) for the axial charges, we discuss the generalization to weak processes in nuclei of the Goldberger-Treiman (G.-T.) relation, the Adler-Weisberger (A.-W.) sum rule, and the Weinberg-Tomozawa (W.-T.) formula and treat in particular the extent of pion-pole dominance of the axial-current divergence form factor in nuclear beta decay.

We consider the nuclear beta decay

$$N_i \to N_f + e + \nu_e \tag{58}$$

where, for the sake of simplicity, we work out explicitly only the case where the spin and isospin of N_i and N_f are the same as the spin and isospin of neutron and proton (e.g. $|N_i\rangle$ and $|N_f\rangle$ are the ground states of H^3 and He^3 or of N^{13} and C^{13}). However, all of our results (i.e. Eqs. (61)–(67) and (69)–(85) below) also hold in the case when the spins of N_i and N_f are $>\frac{1}{2}$ (e.g. $|N_i\rangle$ and $|N_f\rangle$ are the ground states of C^{11} and B^{11} which have spin 3/2 and isospin 1/2) and some (e.g. Eq. (69) below) are valid even when neither the spin nor isospin of N_i and N_f are 1/2 (e.g. $|N_i\rangle$ and $|N_f\rangle$ are the ground states of C^{14} and N^{14} which have, respectively, spin and isospin 0 and 1, and spin and isospin 1 and 0).

Thus, introducing the pseudoscalar form factor as well as the axial form factor, we have [18]

$$\langle N_f|A_\lambda(0)|N_i\rangle = \bar{u}_f\tau^{(-)}\left[\gamma_\lambda\gamma_5 F_A(q^2;N_i\to N_f)+\frac{iq_\lambda(m_i+m_f)}{m_\pi^2+q^2}\gamma_5 F_P(q^2;N_i\to N_f)\right]u_i\,, \tag{59}$$

so that the axial-current divergence form factor is

$$\begin{aligned} D_A(q^2;N_i\to N_f) &\equiv \frac{\langle N_f|\dfrac{\partial A_\lambda(0)}{\partial x_\lambda}|N_i\rangle}{(\bar{u}_f\tau^{(-)}\gamma_5 u_i)\,(m_i+m_f)} \\ &= F_A(q^2;N_i\to N_f)+\frac{q^2}{m_\pi^2+q^2}F_P(q^2;N_i\to N_f)\,. \end{aligned} \tag{60}$$

Then, assuming that $D_A(q^2;N_i\to N_f)$ vanishes as large q^2, which is our version of PCAC in the present context, we can write the unsubtracted dispersion relation

$$D_A(q^2;N_i\to N_f)=\frac{a_\pi f_{\pi N_i N_f}(-m_\pi^2)}{1+q^2/m_\pi^2}+\frac{1}{\pi}\int\limits_{m_0^2}^{\infty}\frac{\operatorname{Im}D_A(-m^2;N_i\to N_f)}{q^2+m^2}\,d(m^2) \tag{61}$$

where we have isolated the pion-pole term. In Eq. (61), we have, from the observed rate of $\pi\to\mu+\nu_\mu$,

$$a_\pi\equiv F_A(-m_\pi^2;\text{pion}\to\text{vacuum})=0.94\pm0.01 \tag{62}$$

while $f_{\pi N_i N_f}(-m_\pi^2)$ is the pion-nucleus-nucleus coupling constant [19]; $f_{\pi N_i N_f}(-m_\pi^2)$ is known only in the case $N_i=n$, $N_f=p$ from the analysis of pion-proton elastic and neutron-proton charge-exchange scattering – this yields

$$\begin{gathered} \frac{1}{2}\,\frac{[f_{\pi np}(-m_\pi^2)]^2}{4\pi}=\frac{[f_{\pi^0 pp}(-m_\pi^2)]^2}{4\pi}=0.081\pm0.003\,, \\ f_{\pi np}(-m_\pi^2)=1.43\pm0.03\,. \end{gathered} \tag{63}$$

It is also to be noted that, due to the presence of anomalous thresholds in the nuclear case, m_0^2 is appreciably less than $(3m_\pi)^2$.

We now introduce the quantity $K_{\pi N_i N_f}(q^2)$, the so-called pionic form factor of the process $N_i \to N_f + \pi$, by the equation

$$D_A(q^2; N_i \to N_f) = \frac{a_\pi f_{\pi N_i N_f}(-m_\pi^2)\, K_{\pi N_i N_f}(q^2)}{1 + q^2/m_\pi^2} \tag{64}$$

so that combining Eqs. (61) and (64)

$$\begin{aligned} K_{\pi N_i N_f}(q^2) &= 1 + \left(\frac{1+q^2/m_\pi^2}{a_\pi f_{\pi N_i N_f}(-m_\pi^2)}\right) \frac{1}{\pi} \int_{m_0^2}^{\infty} \frac{\operatorname{Im} D_A(-m^2; N_i \to N_f)}{q^2 + m^2}\, \mathrm{d}(m^2) \\ &= 1 + (1 + q^2/m_\pi^2) \frac{1}{\pi} \int_{m_0^2}^{\infty} \frac{\operatorname{Im} K_{\pi N_i N_f}(-m^2)}{(q^2+m^2)(1-m^2/m_\pi^2)}\, \mathrm{d}(m^2), \end{aligned} \tag{65}$$

$$K_{\pi N_i N_f}(-m_\pi^2) = 1\,.$$

Thus, pion-pole dominance of the axial-current divergence form factor $D_A(q^2; N_i \to N_f)$ at $q^2 \cong 0$ corresponds to

$$K_{\pi N_i N_f}(0) \cong 1 \tag{66}$$

which is satisfied providing that

$$\left| \frac{1}{\pi} \int_{m_0^2}^{\infty} \frac{\operatorname{Im} K_{\pi N_i N_f}(-m^2)}{m^2(1-m^2/m_\pi^2)}\, \mathrm{d}(m^2) \right| \ll 1\,. \tag{67}$$

No investigation of the validity of Eq. (67) has even been attempted in the nuclear case.

We now obtain the G.-T. relation at any q^2 by combining Eqs. (60) and (64), viz.:

$$F_A(q^2; N_i \to N_f) + \frac{q^2}{m_\pi^2 + q^2} F_P(q^2; N_i \to N_f) = \frac{a_\pi f_{\pi N_i N_f}(-m_\pi^2)\, K_{\pi N_i N_f}(q^2)}{1 + q^2/m_\pi^2} \tag{68}$$

so that, considering the case $q^2 = 0$,

$$F_A(0; N_i \to N_f) = a_\pi f_{\pi N_i N_f}(-m_\pi^2)\, K_{\pi N_i N_f}(0) \tag{69}$$

which is the conventional G.-T. relation. In the case $N_i = n$, $N_f = p$, we have, substituting Eqs. (3), (62), and (63) into Eq. (69),

$$K_{\pi n p}(0) = 0.92 \pm 0.02 \tag{70}$$

so that pion-pole dominance of the axial-current divergence form factor holds to within 8% in the nucleon case. In the nuclear case the appropriate $n + N_f \to p + N_i$ charge-exchange scattering experiments to determine the $f_{\pi N_i N_f}(-m_\pi^2)$ [19] have not yet been performed so that we have no direct information regarding the $f_{\pi N_i N_f}(-m_\pi^2)$; on the other hand, the corresponding $F_A(0; N_i \to N_f)$ are well known from the rate and energy release of the $N_i \to N_f + e + \nu_e$ beta decays (see, e.g., Eq. (2) for $N_i = H^3$, $N_f = He^3$). We therefore need another relation connecting $K_{\pi N_i N_f}(0)$ to measurable quantities in order to test the extent of pion-pole dominance of $D_A(0; N_i \to N_f)$.

Such a relation can be deduced by considering the generalization to nuclei of the A.-W. sum rule and the Goldberger-Miazawa-Oehme (G.-M.-O.) sum rule. We obtain the A.-W. sum rule by using ETCR for the axial charges and the G.-M.-O. sum rule by a straightforward dispersion-theoretic approach [19], viz.:

$$\sum_l [F_A(0; N_l \to N_f)]^2 \kappa_l = \xi_f \tag{71}$$

and

$$\sum_l [f_{\pi N_l N_f}(-m_\pi^2)]^2 \kappa_l = 2\pi \left(1 + \frac{m_\pi}{m_f}\right) (A(\pi^-, N_f) - A(\pi^+, N_f)) \tag{72}$$

where

$$\kappa_l \equiv \kappa(J_l^{(P)}, J_f^{(P)}) (Z_f - Z_l), \qquad \xi_f \equiv (Z_f - (A_f - Z_f)) \tag{73}$$

with

$$\kappa\left(\frac{1^{(\pm)}}{2}, \frac{1^{(\pm)}}{2}\right) = 1, \quad \kappa\left(\frac{3^{(\pm)}}{2}, \frac{1^{(\pm)}}{2}\right) = 2\kappa\left(\frac{1^{(\pm)}}{2}, \frac{3^{(\pm)}}{2}\right) = \frac{2}{3}\left(\frac{\mathrm{m}_l + \mathrm{m}_f}{2m_l}\right)^2,$$
$$\kappa\left(\frac{3^{(\pm)}}{2}, \frac{3^{(\pm)}}{2}\right) = \frac{5}{9}, \ldots, \tag{74}$$
$$Z_f - Z_l = +1 \quad \text{or} \quad -1,$$
$$\xi_f = \begin{matrix} +1, \text{e.g. for } N_f = \mathrm{He}^3 \\ -1, \text{e.g. for } N_f = \mathrm{C}^{13} \end{matrix}$$

and where the $A(\pi^\mp, N_f)$ in Eq. (72) is the real part of the $\pi^\mp - N_f$ scattering length (in units of m_π^{-1}) i.e. the real part of the $\pi^\mp - N_f$ forward elastic scattering amplitude for pions of zero kinetic energy. The sums over l in Eqs. (71) and (72) contain a term from the state $|N_l; Z_l = Z_i$, $m_l = m_i$, $J_l^{(P)} = J_i^{(P)}\rangle$ as well as terms from states $|N_l; Z_l = Z_f \mp 1, m_l > m_i$, $J_l^{(P)}\rangle$ both below $(m_l < m_f + m_\pi)$ and above $(m_l > m_f + m_\pi)$ the threshold for $N_l \to N_f + \pi$. It is especially to be noted that the same "kinematic coefficient" κ_l appears in both the A.-W. sum rule and the G.-M.-O. sum rule; this is a consequence of the fact that the divergence of the axial hadronic weak current has the same transformation properties under space-rotation, space-inversion, and isospace-rotation as the pion field.

We proceed to substitute Eq. (69) into Eq. (71) and then divide Eq. (71) by Eq. (72). This gives

$$\frac{a_\pi^2 \sum_l [f_{\pi N_l N_f}(-m_\pi^2)]^2 [K_{\pi N_l N_f}(0)]^2 \kappa_l}{\sum_l [f_{\pi N_l N_f}(-m_\pi^2)]^2 \kappa_l} = \frac{\xi_f}{2\pi\left(1 + \frac{m_\pi}{m_f}\right)(A(\pi^-, N_f) - A(\pi^+, N_f))} \tag{75}$$

and defining

$$\langle K_{\pi N_l N_f}(0)\rangle^2 \equiv \frac{\sum_l [f_{\pi N_l N_f}(-m_\pi^2)]^2 \, [K_{\pi N_l N_f}(0)]^2 \, \kappa_l}{\sum_l [f_{\pi N_l N_f}(-m_\pi^2)]^2 \, \kappa_l} \tag{76}$$

we obtain

$$\langle K_{\pi N_l N_f}(0)\rangle = \left(\frac{1}{a_\pi}\right)\left[\frac{\xi_f}{2\pi\left(1+\frac{m_\pi}{m_f}\right)(A(\pi^-, N_f) - A(\pi^+, N_f))}\right]^{1/2}. \tag{77}$$

It remains to make the only approximation of our treatment, viz.:

$$\langle K_{\pi N_l N_f}(0)\rangle \cong K_{\pi N_i N_f}(0) \tag{78}$$

which corresponds to the assumption that the dispersion integral in Eq. (65) depends most sensitively on the nuclear mass number $A_l = A_i = A_f$ and much less sensitively on the values of the charge, energy, spin and parity which distinguish among the various states $|N_l; Z_l, m_l, J_l^{(P)}\rangle$. Combination of Eqs. (78) and (77) gives our final result

$$K_{\pi N_i N_f}(0) \cong \left(\frac{1}{a_\pi}\right)\left[\frac{\xi_f}{2\pi\left(1+\frac{m_\pi}{m_f}\right)(A(\pi^-, N_f) - A(\pi^+, N_f))}\right]^{1/2} \tag{79}$$

which is the desired additional relation connecting $K_{\pi N_i N_f}(0)$ to measurable quantities. Eq. (79) can be rewritten in the form

$$\left(\frac{1}{\xi_f}\right)\left(1+\frac{m_\pi}{m_f}\right)(A(\pi^-, N_f) - A(\pi^+, N_f)) \cong \left(\frac{1}{2\pi}\right)\left(\frac{1}{a_\pi^2 [K_{\pi N_i N_f}(0)]^2}\right) \tag{80}$$

or, using Eq. (69) again,

$$\left(\frac{1}{\xi_f}\right)\left(1+\frac{m_\pi}{m_f}\right)(A(\pi^-, N_f) - A(\pi^+, N_f)) \cong \left(\frac{1}{2\pi}\right)\frac{[f_{\pi N_i N_f}(-m_\pi^2)]^2}{[F_A(0; N_i \to N_f)]^2}. \tag{81}$$

Eq. (80) or Eq. (81) constitutes our generalization of the W.-T. formula to nuclei [19]; in the original W.-T. formula as applied to nuclei

$$[f_{\pi n p}(-m_\pi^2)]^2/[F_A(0; n \to p)]^2$$

appeared instead of $[f_{\pi N_i N_f}(-m_\pi^2)]^2/[F_A(0; N_i \to N_f)]^2$, or equivalently, $[K_{\pi n p}(0)]^2$ appeared instead of $[K_{\pi N_i N_f}(0)]^2$. Eq. (80) or Eq. (81) is in excellent agreement with experiment in the case $N_f = p$, $N_i = n$ where we have

$$\left(\frac{1}{\xi_p}\right)\left(1+\frac{m_\pi}{m_p}\right)(A(\pi^-,p)-A(\pi^+,p))$$
$$= (1)\,(1.15)\,(0.188\pm 0.008) = 0.216\pm 0.009\,,$$
$$\left(\frac{1}{2\pi}\right)\left(\frac{1}{a_\pi^2[K_{\pi np}(0)]^2}\right) = \left(\frac{1}{2\pi}\right)\frac{[f_{\pi np}(-m_\pi^2)]^2}{[F_A(0;n\to p)]^2} = 0.215\pm 0.008\,. \quad (82)$$

Further, some rather crude experimental information is available on $(A(\pi^-,N_f)-A(\pi^+,N_f)) = (A(\pi^-,N_f)-A(\pi^-,N_i))$ for $[N_f,N_i] = [He^3, H^3]$, $[Li^7, Be^7]$, $[Be^9, B^9]$, $[B^{11}, C^{11}]$, $[F^{19}, Ne^{19}]$, $[Na^{23}, Mg^{23}]$, the $A(\pi^-,N_f)$ being obtained from measured $2p-1s$ transition energies of the corresponding pionic atoms and the $A(\pi^-,N_i)$ from interpolation among such $2p-1s$ transition energies. This information can be roughly summarized as [20]

$$\left(\frac{1}{\xi_f}\right)\left(1+\frac{m_\pi}{m_f}\right)(A(\pi^-,N_f)-A(\pi^+,N_f)) = 0.11\pm 0.02 \quad (83)$$

so that, using Eq. (79),

$$K_{\pi N_i N_f}(0) \cong 1.3\pm 0.1\,. \quad (84)$$

The result in Eq. (84) seems to indicate that a larger deviation from pion-pole dominance of the axial-current divergence form factor is present in nuclei, namely 30%, than in nucleons (8%: Eq. (70)); moreover the sign of $(K_{\pi N_i N_f}(0)-1)$ is opposite to that of $(K_{\pi np}(0)-1)$ and this qualitative difference may arise from the anomalous threshold contribution $\left(\int_{m_0^2}^{(3m_\pi)^2}\ldots\right)$ to the dispersion integral in Eq. (65). Clearly, a careful study of this dispersion integral as well as precise experimental data on the pion-nucleus scattering lengths would be of great interest.

In conclusion we note that if we rewrite Eq. (81) in the form

$$F_A(0;N_i\to N_f) \cong f_{\pi N_i N_f}(-m_\pi^2)\left[\frac{\xi_f}{2\pi\left(1+\frac{m_\pi}{m_f}\right)(A(\pi^-,N_f)-A(\pi^+,N_f))}\right]^{1/2} \quad (85)$$

we have a relation of the G.-T. type with all quantities entering being directly measurable. Experimental verification of Eq. (85) for a variety of nuclei, which would require accurate determination of the $f_{\pi N_i N_f}(-m_\pi^2)$ as well as the $A(\pi^\mp, N_f)$, would provide valuable evidence for PCAC (required for the expressions for $D_A(q^2;N_i\to N_f)$ and $K_{\pi N_i N_f}(q^2)$ in Eqs. (61) and (65)), for ETCR (required for the A.-W. rule in Eq. (71)), and for our approximation for $\langle K_{\pi N_i N_f}(0)\rangle$ (Eq. (78)).

III. Nonleptonic Weak Processes in Nuclei — Space-Inversion Noninvariant Weak Nuclear Forces

Nonleptonic weak processes in nuclei give rise to space-inversion non invariant internucleon potentials, $V_{\text{weak nucl.}}$, which induce "wrong-parity" admixtures into the various nuclear states:

$$\Psi_P \to \sqrt{1-|F|^2}\,\Psi_P + F\Psi_{-P}\,,$$
$$F = \frac{\langle \Psi_{-P} | V_{\text{weak nucl.}} | \Psi_P \rangle}{(E_P^{(0)} - E_{-P}^{(0)})}\,. \tag{86}$$

The potentials $V_{\text{weak nucl.}}$ can be viewed as arising predominantly from the exchange of a π-meson and a ϱ-meson between the two nucleons [21]

$$V_{\text{weak nucl.}}(\boldsymbol{r}_{12};\boldsymbol{\sigma}_1,\boldsymbol{\sigma}_2;\boldsymbol{\tau}_1,\boldsymbol{\tau}_2) = V^{(\pi)}_{\text{weak nucl.}}(\boldsymbol{r}_{12};\boldsymbol{\sigma}_1,\boldsymbol{\sigma}_2;\boldsymbol{\tau}_1,\boldsymbol{\tau}_2) + V^{(\varrho)}_{\text{weak nucl.}}(\boldsymbol{r}_{12};\boldsymbol{\sigma}_1,\boldsymbol{\sigma}_2;\boldsymbol{\tau}_1,\boldsymbol{\tau}_2)\,,$$

$$\begin{aligned} V^{(\pi)}_{\text{weak nucl.}}(\boldsymbol{r}_{12};\boldsymbol{\sigma}_1,\boldsymbol{\sigma}_2;\boldsymbol{\tau}_1,\boldsymbol{\tau}_2) &\approx (G_\mu \sin^2\vartheta m_\pi^2)(f_{\pi n p})(\boldsymbol{\sigma}_1+\boldsymbol{\sigma}_2) \\ &\quad \cdot\left(\frac{\boldsymbol{r}_{12}}{r_{12}}\right)\left(\frac{\mathrm{e}^{-m_\pi r_{12}}}{r_{12}}\right)(\tau_1^{(-)}\tau_2^{(+)} - \tau_1^{(-)}\tau_2^{(-)}) \\ &\approx 10^{-8}\, m_\pi \quad (\text{at } r_{12} = m_\pi^{-1})\,, \end{aligned} \tag{87}$$

$$\begin{aligned} V^{(\varrho)}_{\text{weak nucl.}}(\boldsymbol{r}_{12};\boldsymbol{\sigma}_1,\boldsymbol{\sigma}_2;\boldsymbol{\tau}_1,\boldsymbol{\tau}_2) &\approx (G_\mu \cos^2\vartheta m_\varrho^2)(f_{\varrho n p})(\boldsymbol{\sigma}_1\times\boldsymbol{\sigma}_2) \\ &\quad \cdot\left(\frac{\boldsymbol{r}_{12}}{r_{12}}\right)\left(\frac{\mathrm{e}^{-m_\varrho r_{12}}}{r_{12}}\right)(\tau_1^{(-)}\tau_2^{(+)} + \tau_1^{(+)}\tau_2^{(-)}) \\ &\approx 10^{-7}\, m_\pi \quad (\text{at } r_{12} = m_\pi^{-1}) \end{aligned}$$

one of the meson-nucleon-nucleon vertices being weak and the other strong. Eq. (87) shows that $V^{(\pi)}_{\text{weak nucl.}}$ transforms like an isovector and $V^{(\varrho)}_{\text{weak nucl.}}$ like a linear combination of an isoscalar and an isotensor. The weak meson-nucleon-nucleon coupling constants depend on whether the nonleptonic strangeness-preserving weak Hamiltonian, $H^{(\Delta S=0)}_{\text{weak n.l.}}$, contains *neutral* hadronic weak currents, the sensitivity of $V^{(\pi)}_{\text{weak nucl.}}$ to the presence of these currents being considerably greater than that of $V^{(\varrho)}_{\text{weak nucl.}}$ – we note that the weak meson-nucleon-nucleon coupling constants set down in Eq. (87), $G_\mu \sin^2\vartheta m_\pi^2$ and $G_\mu \cos^2\vartheta m_\varrho^2$, are appropriate to the case where the neutral hadronic weak currents are absent.

The above discussion shows that comparison of theoretical and experimental values of the wrong-parity admixture coefficients F, the former calculated from $V^{(\pi)}_{\text{weak nucl.}}$ and $V^{(\varrho)}_{\text{weak nucl.}}$ appropriate to given

$H^{(\Delta S=0)}_{\text{weak n.l.}}$ plus suitable nuclear physics, and the latter extracted from relevant measurements of γ-ray circular polarizations and asymmetries, can distinguish among various assumptions regarding $H^{(\Delta S=0)}_{\text{weak n.l.}}$. Available studies of this kind seem to favor $H^{(\Delta S=0)}_{\text{weak n.l.}}$ *without* neutral hadronic weak currents; however, many additional measurements and calculations will be necessary before this conclusion can be considered at all definite [21].

We conclude our very brief discussion of the $V_{\text{weak nucl.}}$ by noting that Lobashov has very recently described an ingenious experiment on the circular polarization of the γ-ray in $n+p \to d+\gamma$ with preliminary results which appear to give a circular polarization $P_\gamma \approx 10^{-6}$ [22]; this value is not inconsistent with a prediction made on the basis of Eqs. (87) and (86) $\left(P_\gamma \approx F \approx \dfrac{10^{-7} m_\pi}{(5\ \text{Mev})} = 3\times 10^{-6}\right)$. It is clear that the $n+p\to d+\gamma$ experiment offers the greatest promise for a really fundamental investigation of $V_{\text{weak nucl.}}$ since the calculation of F is relatively simple when only two nucleons are involved; also, the γ-ray circular polarization in this case can arise only from the isoscalar part of $V_{\text{weak nucl.}}$, i.e. from $V^{(\varrho)}_{\text{weak nucl.}}$, and this is more difficult to calculate reliably and therefore more interesting than $V^{(\pi)}_{\text{weak nucl.}}$.

References and Remarks

1. CVC ≡ **C**onserved **V**ector hadronic weak **C**urrent hypothesis which identifies this current with the hadronic isospin current. Eq. (1) then follows since $|H^3\rangle$ is transformed into $|He^3\rangle$ by the charge-symmetry operator $\exp(-i\pi I^{(2)})$ – see, e.g., *Kim, C. W., Primakoff, H.*: Phys. Rev. **139**, B 1447 (1965).
2. One has:

$$\sum_{M_f=-J}^{J} \left|\langle \Psi_{M_f}(\text{He}^3)| \sum_{k=1}^{3} \tau_k^{(+)}\sigma_k |\Psi_{M_i}(\text{H}^3)\rangle\right|^2$$
$$= 3\left\{|a(^2S_{1/2})|^2 - \frac{1}{3}|a(^2S'_{1/2})|^2 + \frac{1}{3}|a(^4D_{1/2})|^2 + \cdots\right\}^2$$
$$= 3\left\{1 - \frac{4}{3}|a(^2S'_{1/2})|^2 - \frac{2}{3}|a(^4D_{1/2})|^2 + \cdots\right\}^2$$

where

$$\Psi_{M_i}(\text{H}^3) = a(^2S_{1/2})\,\Psi_{M_i}(\text{H}^3;{}^2S_{1/2}) + a(^2S'_{1/2})\,\Psi_{M_i}(\text{H}^3;{}^2S'_{1/2}) + a(^4D_{1/2})\,\Psi_{M_i}(\text{H}^3;{}^4D_{1/2}) + \cdots$$

and similarly for $\Psi_{M_f}(\text{He}^3)$ – see *Blatt, J. M.*: Phys. Rev. **89**, 86 (1953). According to *Gibson* (see *Gibson, B.*: Phys. Rev. **139**, B 1153 (1953) and Nucl. Phys. B **2**, 501 (1967)) optimum values for the $|a|^2$ are: $|a(^2S_{1/2})|^2 = 0.92$, $|a(^2S'_{1/2})|^2 = 0.02$, $|a(^4D_{1/2})|^2 = 0.06$, all other $|a|^2$ very small. See also *Delves, L. M., Blatt, J. M., Pask, C., Davies, B.*: Phys. Letters B **28**, 472 (1969).

3. One has:

$$[\mu(\mathrm{H}^3)-\mu(\mathrm{He}^3)]_{n-p\ \mathrm{nuc.\,mod.}}$$

$$=(\mu(p)-\mu(n))\left\{|a(^2S_{1/2})|^2-\frac{1}{3}|a(^2S'_{1/2})|^2+\frac{1}{3}(1-1/(\mu(p)-\mu(n)))\,|a(^4D_{1/2})|^2+\cdots\right\}$$

$$=(\mu(p)-\mu(n))\left\{1-\frac{4}{3}|a(^2S'_{1/2})|^2-\frac{2}{3}(1+1/2(\mu(p)-\mu(n)))\,|a(^4D_{1/2})|^2+\cdots\right\}$$

where $\mu(p)=2.79\left(\frac{e}{2m_p}\right)$ and $\mu(n)=-1.91\left(\frac{e}{2m_p}\right)$ are the measured magnetic moments of the proton and the neutron – see *Verde, M.*: Handbuch der Physik **39**, 144 (1957). With the $|a|^2$ of Ref. [2] this gives

$$[\mu(\mathrm{H}^3)-\mu(\mathrm{He}^3)]_{n-p\ \mathrm{nuc.\,mod.}}=(\mu(p)-\mu(n))\,(0.93\pm0.02)=(4.37\pm0.08)\left(\frac{e}{2m_p}\right).$$

4. PCAC ≡ **P**artially **C**onserved **A**xial hadronic weak **C**urrent hypothesis applied in the present context in the manner described in *Adler, S. L.*: Ann. Phys. (N. Y.) **50**, 189 (1968). See also Section II d) of this paper.
5. See *Cheng, W. K.*, to be published, and *Chemtob, M., Rho, M.*: Phys. Letters B **29**, 540 (1969).
6. The only nonvanishing matrix elements of $t^{(+)}$ are:

$$\langle p|\,t^{(+)}\,|n\rangle=1\,,$$

$$\langle p|\,t^{(+)}\,|N^{*0}\rangle=\langle N^{*+}|\,t^{(+)}\,|n\rangle=\frac{F_A(q^2=0;N^{*0}\to p)}{F_A(q^2=0;n\to p)}\,.$$

$$\langle N^{*+}|\,t^{(+)}\,|N^{*0}\rangle=\frac{F_A(q^2=0;N^{*0}\to N^{*+})}{F_A(q^2=0;n\to p)}$$

while the only nonvanishing matrix element of $\tau^{(+)}$ is:

$$\langle p|\,\tau^{(+)}\,|n\rangle=1\,.$$

7. For a general discussion of the Goldberger-Treiman relation see, e.g., Ref. [1], and also section II d) of this paper. The numerical value of f^* is found from the partial width of the $N^{*+}(1470)\to p+\pi^0$ process, viz.: 140 Mev.
8. We note that Eq. (13) predicts $C^2\approx0.02(f^*/f)^2=0.3\%$; this is to be compared with a calculation of *Kerman, A. K., Kisslinger, L. S.*: Phys. Rev. **180**, 1483 (1969), who estimate that the probability of finding the nucleon-isobar $N^*(1688$ Mev; $J=5/2$, $I=1/2)$ within the deuteron is $1/2\%-1\%$. The Kerman-Kisslinger estimate is made in the course of an analysis of the quantitative inadequacy of the conventional neutron-proton nuclear model in the description of high-energy backward elastic proton-deuteron scattering. See also *Arenhovel, H., Danos, M.*: Phys. Letters B **28**, 299 (1968), and *Kisslinger, L.*: Phys. Letters B **29**, 211 (1969).
9. The absolute square of the $I=1$ and $I=2$ impurity in $\mathrm{He}^4(J=0^{(+)},I=0)$ is no more than a small fraction of a percent while the absolute square of the $I=1$ impurity in $\mathrm{H}^4(J=0^{(+)},I=2)$ is really tiny since there is no internucleon Coulomb force in H^4.
10. See *Davis, R.*: Phys. Rev. **97**, 766 (1955) for the non-observation of $\nu_e+n_{\mathrm{bound}}\to e^-+p_{\mathrm{bound}}$ and *Borer, K., Hahn, B., Hofer, H., Kaspar, H., Krienen, F., Seiler, P. G.*: Phys. Letters B **29**, 615 (1969) for the non-observation of $\nu_\mu+p_{\mathrm{bound}}\to\mu^++n_{\mathrm{bound}}$.
11. *Kirsten, T., Schaeffer, O. A., Norton, E., Stoenner, R. W.*: Phys. Rev. Letters **20**, 1300 (1968); *Kirsten, T., Gentner, W., Schaeffer, O. A.*: Z. Physik **202**, 273 (1967); *Kirsten, T., Muller, H. W.*: To be published.

12. *Takaoka, N., Ogata, K.*: Z. Naturforsch. **21a**, 84 (1966).
13. We note that the reaction $Te^{127}+\alpha\to Xe^{130}+n$ cannot take place in Te rocks since Te^{127} is unstable to single-beta decay with a lifetime of 9.4 hours.
14. *Primakoff, H., Rosen, S. P.*: Phys. Rev. **184**, 1925 (1969).
15. *Primakoff, H., Sharp, D. H.*: Phys. Rev. Letters **23**, 501 (1969).
16. *Kim, C. W., Primakoff, H.*: Phys. Rev. **180**, 1502 (1969).
17. In the case $Ne^{19}\to F^{19}$, the operators $I_0^{(+)}(0)$, $\{A_3(0)\}^\dagger_{CP\text{-normal}}$, and $\{A_3(0)\}^\dagger_{CP\text{-abnormal}}$ in the $\langle\beta|\ldots|\alpha\rangle$ matrix elements must be replaced by the corresponding hermitian-conjugate operators.
18. Here and in what follows we take $A_\lambda(x)=\{A_\lambda(x)\}_{CP\text{-normal}}$.
19. *Primakoff, H.*: High energy physics and nuclear structure, pp. 409–466. *Alexander, G.*, Ed. Amsterdam: North-Holland Publishing Company 1967.
20. *Seki, R.*: Phys. Letters **23**, 1000 (1969).
21. See the review by *McKellar, B. H. J.*: Proceedings of Third International Conference on High Energy Physics and Nuclear Structure. *Devons, S.*, Ed. New York: Plenum Press Publishing Company 1969.
22. *Lobashov, V. M.*: Proceedings of Third International Conference on High Energy Physics and Nuclear Structure. *Devons, S.*, Ed. New York: Plenum Press Publishing Company 1969.

Prof. Dr. *H. Primakoff*
Department of Physics
University of Pennsylvania
Philadelphia, Pennsylvania 19104/USA

Weak Interactions at High Energies

G. v. GEHLEN

Contents

1. Introduction

All weak interaction processes which have been studied experimentally are compatible with the assumption that the weak interaction can be described by the interaction hamiltonian density

$$\mathscr{H}_I(x) = \frac{G}{\sqrt{2}} j_\mu(x)\, j_\mu^+(x) \tag{1}$$

where $j_\mu(x)$ is the weak current consisting of a leptonic part j_μ^l and strangeness conserving and strangeness violating hadronic parts j_μ^h:

$$j_\mu = j_\mu^l + j_\mu^h . \tag{2}$$

The coupling constant G is known from the decay $\mu \to e + \nu + \bar{\nu}$ to be

$$G = 1.01 \cdot 10^{-5}\, m_N^{-2} \tag{3}$$

where m_N is the nucleon mass.

The hamiltonian (1) is understood to be an effective hamiltonian which has to be used in lowest order perturbation theory. Only the $K_L - K_S$-mass difference is supposed to be a second order effect of (1). Whether the observed CP-violation effects in K-decay are to be explained by an interaction like (1), is still unclear. Of course, in many processes, effects of strong interactions which cannot be calculated in detail, make it impossible to draw definite conclusions from (1), but in processes not involving neutral K-mesons there is no evidence against the assumption that the weak interactions are described by (1), evaluated in lowest order.

In the following lectures, we shall consider weak reactions by which deviations from (1) may be detected. In the first part we study the possibility that (1) is the limiting case of a second order effect of the hamiltonian

$$\mathscr{H}_I(x) = g j_\mu(x)\, W_\mu(x) + \text{h.c.} \tag{4}$$

where W_μ is the field of an "intermediate vector boson". Since the problem of assigning $SU(3)$-properties to the particles W will be treated in detail in the lectures of Prof. *G. Segrè*, we shall not enter into these details and just assume a pair of charged bosons W^+ and W^-.

Both hamiltonians (1) and (4) lead to divergences if we try to evaluate them in higher order perturbation theory. However, one may think that for processes involving hadrons the presence of strong interaction form factors may render higher order calculations of (1) or (4) finite. In the second part of our lectures we shall show, following the work of *Ioffe* and *Shabalin* [1] and *Mohapatra, Rao*, and *Marshak* [2], that if we assume current algebra commutation relations and the validity of Bjorkens method for performing high energy limits [3], even higher order processes of (1) or (4) involving hadrons need a cut-off if we want a finite result.

2. Intermediate Vector Boson Theory

By considering the second order graph of the interaction (4) in which a W is exchanged between two currents j_μ, we easily find the relation between the coupling constants G and g which appear in (1) and (4), respectively. The low q limit of the W-propagator is

$$\frac{\delta_{\mu\nu} + q_\mu q_\nu / m_W^2}{q^2 + m_W^2} \xrightarrow[q_\mu \ll m_W]{} \frac{\delta_{\mu\nu}}{m_W^2}\,. \tag{5}$$

Therefore, for small q_μ, the second order effects of (3) reduce to the contact interaction (1) if

$$\frac{G}{\sqrt{2}} = \frac{g^2}{m_W^2}\,. \tag{6}$$

The maximal momentum transfer appearing in weak decay processes is quite low, therefore the differences between the predictions of (1) and (4) are small. Consider e.g. μ-decay, $\mu \to e + \nu + \bar{\nu}$. Using the universal

Fermi interaction (UFI) of Eq. (1) with pure *V–A*-coupling, the electron spectrum is characterized by the ϱ-value $\varrho = 3/4$, whereas according to the intermediate vector boson theory (IVB) we obtain approximately [4]

$$\varrho \approx \frac{3}{4} + \frac{1}{3}\left(\frac{m_\mu}{m_W}\right)^2 \tag{7}$$

disregarding radiative corrections. Experimentally, a recent result [5] for ϱ is (approx. 5% radiative corrections have been taken into account):

$$\varrho = 0.751 \pm 0.003 . \tag{8}$$

If one believes in exact *V–A* coupling and if one makes the unjustified assumption that the radiative corrections do not change by going from UFI to IVB (see, however, *Bailin* [6]), one would conclude from (7) and (8)

$$m_W \gtrsim 1 \text{ GeV} . \tag{9}$$

However, from the absence of the decay

$$K \to W + \gamma \tag{10}$$

a much more direct conclusion about the mass of the W can be made: (10) is impossible if

$$m_W > m_K .$$

The best limit on m_W which has been obtained up to now comes from the unsuccessful attempts to produce the W-mesons directly. If we assume that the W is coupled to the weak current according to (4), W's can be produced in almost every process if the energy is sufficiently high. However, it will be quite difficult to distinguish the produced W's against other particles. The W must be so short-lived that it can not be observed directly, only via its decay products. From the decay mode

$$W \to \mu + \nu \tag{11}$$

which is possible according to (3), one obtains an upper limit for the decay rate of the W:

$$\Gamma_{W\to\mu+\nu} = \frac{G m_W^3}{6\pi\sqrt{2}} \approx 5.4 \cdot 10^{17} \left(\frac{m_W}{m_N}\right)^3 \sec^{-1} . \tag{12}$$

Other possible decay modes will be

$$\begin{aligned} &W \to e + \nu ,\\ &W \to \pi + \pi ,\\ &W \to K + \gamma \qquad \text{etc.,} \end{aligned}$$

but the complications due to strong interactions make a reliable prediction of the decay rate into hadrons difficult. So we only know

$$\Gamma_{W \to \text{all}} > 10^{18}\ \text{sec}^{-1}\,. \tag{13}$$

3. Production of *W* in Neutrino Experiments

The cleanest experiment for the production of W's is to produce the W in ν-nucleon collisions, because there background effects are small. One looks for the leptonic decay mode (11), i.e. one effectively wants to observe

$$\nu + \text{Nucleus} \to \text{Nucleus} + l^- + l^+ + \nu\,, \tag{14}$$

where l stands for μ or e. The role of the nucleus in (14) is to provide the momentum balance by means of its electromagnetic field. The relevant diagrams are given in Fig. 1a and 1b.

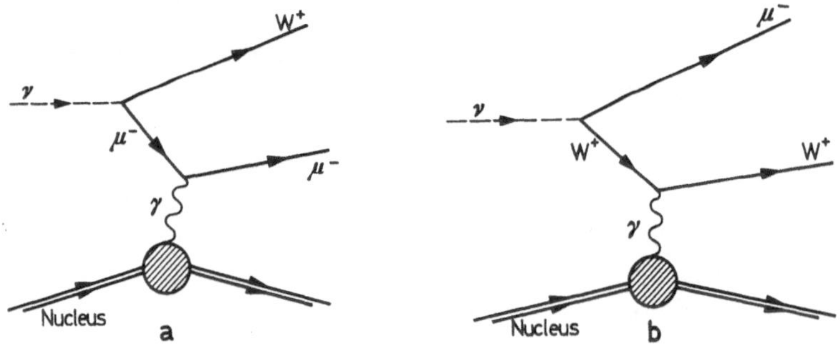

Fig. 1. Diagrams for W-production in neutrino-nucleus collisions

The first diagram, Fig. 1a, can be calculated with only m_W as parameter, whereas the second diagram, Fig. 1b, depends also on the possible anomalous magnetic and anomalous quadrupole moments of the W (in theories without strong interactions of the W the anomalous moments of the W are expected to be small). At low energies and for small anomalous moments of the W the main contribution to the cross section comes from Fig. 1a because the muon is much lighter than the W.

An actual calculation [7–9] has to consider the coherent and incoherent mechanisms for the electromagnetic interaction of the nucleus. In the coherent case the nucleus recoils as a whole while in the incoherent case a single nucleon inside the nucleus interacts with the photon. In the latter case, some nuclear physics enters into the calculation: The Fermi motion of the nucleons gives an important displacement of the threshold, and further, the Pauli principle demands that the momentum transfer has to be so large that the interacting nucleon can leave the nucleus.

Kinematically, at incident neutrino energy E_ν, the momentum tranfer to the nucleus q^2 is restricted by*

$$\frac{m_W^2}{2E_\nu} < \sqrt{q^2} < 2E_\nu \,. \tag{15}$$

The Hofstadter form factor of a nucleus is different from zero only below $q^2 \lesssim 100$ MeV. Therefore, according to (15), the coherent effect contributes strongly only for high E_ν. For low E_ν the coherent contribution furthermore depends only on the badly known tail of the nuclear form factor and cannot be calculated precisely. In contrast, the incoherent effect can contribute strongly at low E_ν due to the slow decrease of the form factor of a single nucleon. Fig. 2 [9] shows results of an explicit evaluation of graphs Fig. 1a and 1b for different values of m_W, assuming the anomalous moments of the W to be zero.

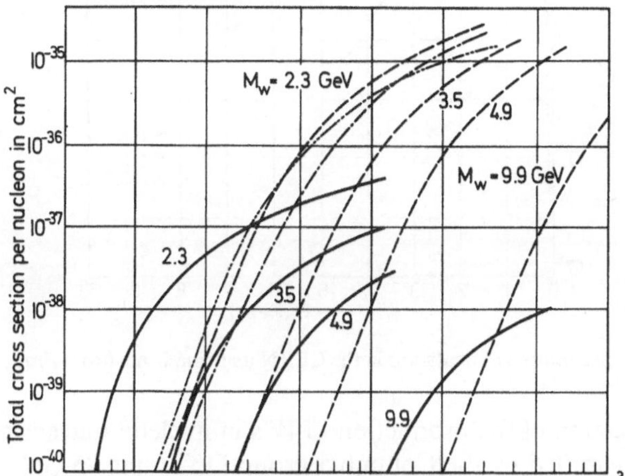

Fig. 2. Total cross section for W^+-production according to the graphs of Fig. 1. Fermi motion and Pauli principle are ignored. The muon mass is neglected in the matrix elements, the W is assumed to have no anomalous moments. The labels on the curves denote the assumed mass of the W. Full lines: Incoherent contribution for a proton. Broken lines (- - -): Coherent cross section per proton for Cu (trapezoidal nuclear form factor). For $M_W = 2.3$ GeV also the coherent cross section for Pb (- · · -) and Al (- · · · -) is shown. The curves give $\frac{1}{13}\,\sigma_{\mathrm{Al}}$ and $\frac{1}{29}\,\sigma_{\mathrm{Cu}}$, but $\frac{1}{207}\,\sigma_{\mathrm{Pb}}$ in order to avoid overlapping of too many curves

* For simplicity, we give in (15) only the approximative formula for the case $m_W \ll E_\nu \ll M_{\mathrm{Target}}$.

Compare this with the fact that the cross section for the quasielastic process [10]

$$\nu + n \to p + \mu^- \tag{16}$$

is almost constant for $E_\nu > 1$ GeV with a value of $\sigma \approx 0.8 \cdot 10^{-38}$ cm^2. One sees that for E_ν high enough, the semiweak reaction (14) will be completely dominant.

Actual neutrino beams at CERN or BNL have an energy distribution concentrated around $E_\nu \approx 1$ GeV, see Fig. 3. Therefore the negative result of the search for *W*-decay [12] excludes only $m_W < 1.8$ GeV. With a future 300 GeV accelerator it is expected that the limit on m_W can be pushed up to 5···8 GeV [10].

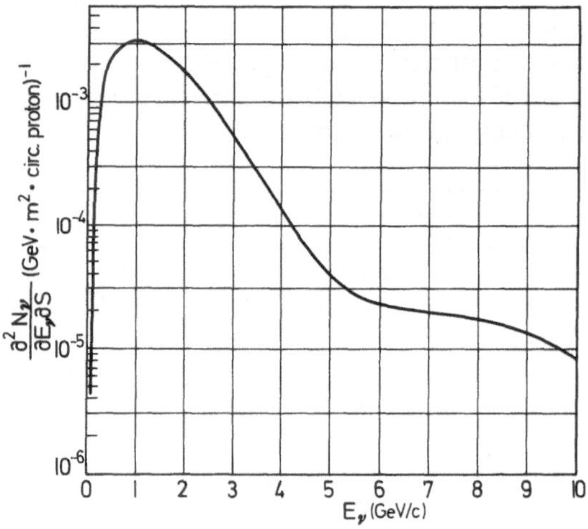

Fig. 3. Energy distribution of neutrinos in the CERN neutrino beam (from *Franzinetti* [11])

The detection of the production of *W*'s in nucleon-nucleus collisions has been attempted by the Columbia group [13] using 25 GeV-protons. Here the main problem is to distinguish the *W*-decay $W \to \mu + \nu$ from an enormous background of muons originating from π-decay. However, the pions which are produced in the *p*-nucleus collisions emerge in a narrow forward cone and their decay muons travel also in the forward direction. The *W*, being heavy, can be produced under large angles, giving rise to muons with large transverse momenta. No clear evidence for such muons under large angles was found, but there are no reliable estimates of the expected *W*-production rate in *p*-nucleus collisions so that the interpretation of the negative result in terms of a limit on the *W*-mass is difficult.

4. Two Neutrino Hypothesis

If we assume that there exist charged W-mesons with the interaction (3), it is necessary to assume that the neutrinos coupled to the muon and electron are different in order to avoid the decay $\mu \to e + \gamma$ proceeding through the graphs of Fig. 4 [14]. Although the evaluation of these graphs is somewhat ambiguous, from Fig. 4 one expects a branching ratio

$$\Gamma(\mu \to e + \gamma)/\Gamma(\mu \to \text{all}) \approx 10^{-4} \cdots 10^{-5}, \tag{17}$$

whereas experimentally this ratio is $< 6 \cdot 10^{-9}$.

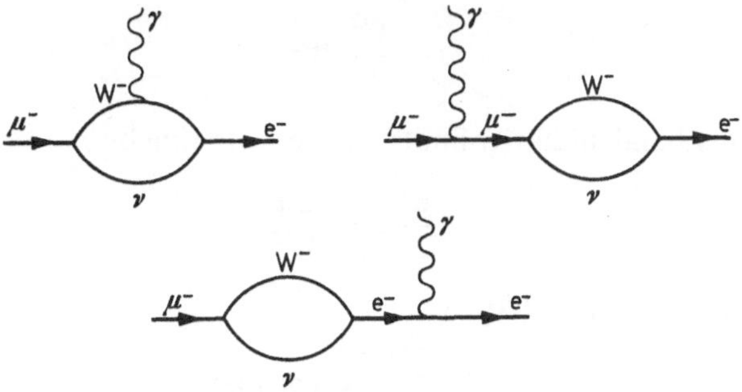

Fig. 4. Graphs leading to the decay $\mu \to e + \gamma$ if W exists and if $\nu_e = \nu_\mu$

More direct evidence for $\nu_e \neq \nu_\mu$ can be obtained from high energy neutrino experiments [15]. The CERN neutrino beam is almost pure ν_μ, because the neutrinos are produced by the decay $\pi^+ \to \mu^+ + \nu_\mu$ and only 0.5–1% contamination of ν_e from $K^+ \to \pi^0 + e^+ + \nu_e$ is present. Therefore, if $\nu_e \neq \nu_\mu$, in the "elastic" reactions

$$\nu_\mu(\nu_e) + n \to \mu^-(e^-) + p \tag{18}$$

mostly muons should be produced with only $\lesssim 1\%$ electron admixture. This is experimentally verified [12] and gives direct evidence that the reaction

$$\nu_\mu + n \to e^- + p$$

is forbidden.

5. Violation of Unitarity in Lowest Order Weak Interactions

In the first sections we discussed possible experiments which may distinguish between UFI, Eq. (1) and the IVB-model, Eq. (4). The predictions of the W-production cross sections and the corresponding limits on the W-mass which were derived from present neutrino experiments were based on the assumption that lowest order perturbation theory with the hamiltonian (4) can be used. However, it is easy to show that at high energies the lowest order approximation of both (1) and (4) leads to conflict with unitarity.

For UFI, consider the elastic scattering process

$$\nu + e \rightarrow \nu + e . \tag{19}$$

The UFI is a pointlike interaction and produces only s-wave scattering. Since no form factors are known which can damp the scattering in (19), the cross section rises proportional to phase space at high center-of-mass energies W:

$$\sigma \sim \frac{G^2}{\pi} W^2 . \tag{20}$$

On the other hand, unitarity limits s-wave scattering by

$$\sigma < \frac{\pi}{2} \lambda^2 = \frac{2\pi}{W^2} , \tag{21}$$

which is in conflict with (20) for

$$W^2 > \sqrt{\frac{2\pi^2}{G^2}} \approx (620 \text{ GeV})^2 . \tag{22}$$

In the IVB-model, the second order cross section for (19) rises at high W only as

$$\sigma \sim G^2 \log W^2 . \tag{23}$$

Thus in IVB unitarity is violated in the process (19) only at much higher energies. However, in the IVB-model there is another process [16]

$$\nu + \bar{\nu} \rightarrow W^+ + W^- \tag{24}$$

which leads into trouble with unitarity at a similar energy like (22). *Gell-Mann et al.* [16] consider a special helicity amplitude for the pair production process (24) and find

$$F^{J=1}_{0,0;-\frac{1}{2},\frac{1}{2}} \rightarrow \frac{GW}{12\pi} \tag{25}$$

for $W \rightarrow \infty$. For this helicity amplitude therefore unitarity is violated if

$$W > \sqrt{\frac{24\pi}{G}} \approx 2600 \text{ GeV} . \tag{26}$$

We see that neither the first order calculation with the hamiltonian (1) nor the second order approximation of (4) can be valid around $W \approx 10^3$

GeV We are left with two possibilities: Either the interaction mechanism for weak interactions is different from (1) and (4), or the inclusion of higher order contributions of (1) or (4) restores unitarity in UFI or IVB theory.

6. Tests of the Locality of the Lepton Interaction

If higher order interactions become important, the leptons in a scattering process

$$\nu + A \rightarrow \mu^- + B \tag{27}$$

(where we assume e.g. A to be a nucleon and B an arbitrary hadron state) no longer interact at one point. This property of the lepton interaction can be tested experimentally (similar as a Rosenbluth-plot tests one-photon exchange in electron scattering) by means of a theorem due *Pais* [17] and *Lee* and *Yang* [18]. The theorem states that the laboratory differential cross section for the process (27) at fixed momentum transfer $q^2 = (p^{(\nu)} - p^{(\mu)})^2$, fixed mass of the created hadron system B and fixed incident energy $E^{(\nu)}$ depends only quadratically on the final muon energy $E^{(\mu)}$:

$$\frac{\mathrm{d}\sigma}{\mathrm{d}q^2\,\mathrm{d}M_B^2} = A + \frac{B}{E^{(\mu)}} + \frac{C}{E^{(\mu)2}} \tag{28}$$

if the leptons interact at one point, see Fig. 5.

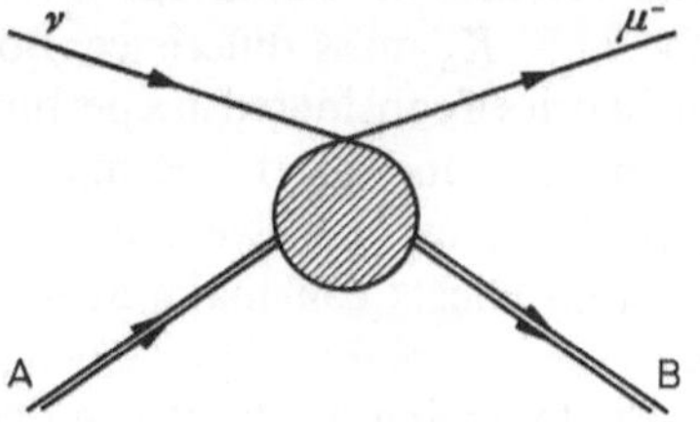

Fig. 5. Diagram for a general inelastic neutrino scattering process

For the proof of (28) we observe that the square of the spin-averaged matrix element for (27) is given by

$$\begin{aligned} |\mathcal{M}|^2 &= \sum_{\text{spin}} |\overline{u}(p^{(\mu)})\, \gamma_\mu (1+\gamma_5)\, u(p^{(\nu)})\, J_\mu|^2 \\ &= (p_\mu^{(\mu)} p_\nu^{(\nu)} + p_\mu^{(\nu)} p_\nu^{(\mu)} - (p^{(\mu)} p^{(\nu)})\, \delta_{\mu\nu} \\ &\quad - \varepsilon_{\mu\nu\varrho\sigma} p_\varrho^{(\mu)} p_\sigma^{(\nu)})\, T_{\mu\nu} \end{aligned} \tag{29}$$

where the hadronic part $T_{\mu\nu}$ can only depend on the transferred momentum $q_\mu = p_\mu^{(\mu)} - p_\mu^{(\nu)}$, but not on $p_\mu^{(\mu)}$ or $p_\mu^{(\nu)}$ separately because of our assumption of a local lepton interaction. So we can write

$$\begin{aligned} T_{\mu\nu} = A_1\, \delta_{\mu\nu} + A_2 (p_\mu^{(A)} p_\nu^{(B)} + p_\mu^{(B)} p_\nu^{(A)}) \\ + A_3 p_\mu^{(A)} p_\nu^{(A)} + A_4 p_\mu^{(B)} p_\nu^{(B)} \end{aligned} \tag{30}$$

where the A_i depend on q^2 and the masses $p^{(A)2}$ and $p^{(B)2} = -M_B^2$. Therefore, for fixed q^2 and M_B, $|\mathcal{M}|^2$ can depend on $E^{(\mu)}$ only through the scalar products $(p^{(\mu)} p^{(\nu)})$, $(p^{(\nu)} p^{(A)})$, $(p^{(\nu)} p^{(B)})$, etc. If we evaluate these scalar products in the laboratory system $\boldsymbol{p}^{(A)} = 0$, we easily find that $|\mathcal{M}|^2$ depends at most quadratically on $E^{(\mu)}$. Inclusion of the phase space and flux factors then leads to (28).

Tests of the general form (28) have not yet been performed. Theorems analogous to (28) for the dependence on $E^{(\nu)}$ and on the cosine of the scattering angle follow by the same argument [17].

7. The Decay $K_L \to \mu^+ + \mu^-$ in Second Order Weak Interaction

The only observed process which probably is due to a second order weak interaction* is the transition $K_0 \leftrightarrow \bar{K}_0$ which leads to the $K_L^0 - K_S^0$ mass difference. Another process which should appear in second order is the decay

$$K_L^0 \to \mu^+ + \mu^- , \tag{31}$$

which is forbidden in first order if no neutral lepton current is present in (2). The question arises whether the hamiltonians (1) or (3) can be used to calculate (31) or the $K_L - K_S$-mass difference. Both interactions lead to nonrenormalizable theories if considered in a perturbation expansion**. However, since both in (31) and in the K-mass difference problem, strongly interacting particles are involved, it is interesting to study whether strong interaction effects can make at least the second order approximation finite.

As an example, in the following we discuss in detail the process (31). In second order of the UFI, (31) can proceed by the mechanism shown in Fig. 6a. Similarly, using a fourth order IWB-interaction, we obtain (31) as shown in Fig. 6b. Both models lead to very similar results. We shall concentrate on Fig. 6a and use the UFI, following the work of *Ioffe* and *Shabalin* [1] and *Mohapatra et al.* [2].

* The situation concerning double β-decay will be discussed in the lectures of Prof. *Primakoff*.

** See the lectures of Prof. *Gatto* for a method of increasing the convergence of the perturbation expansion in IVB-theory.

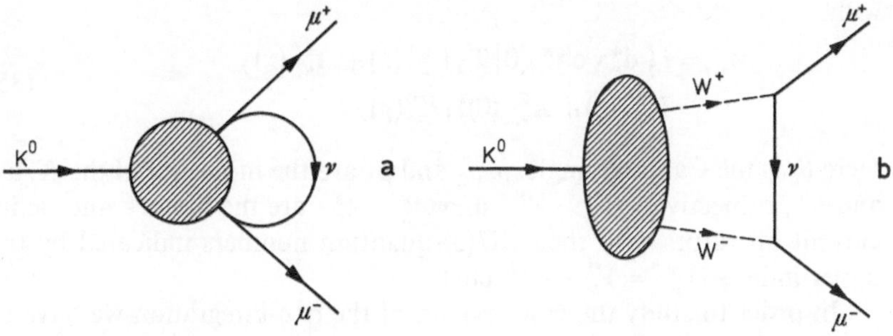

Fig. 6a and b. Diagram for the decay $K_L \to \mu^+ + \mu^-$: a in second order UFI, b with fourth order IVB-interaction

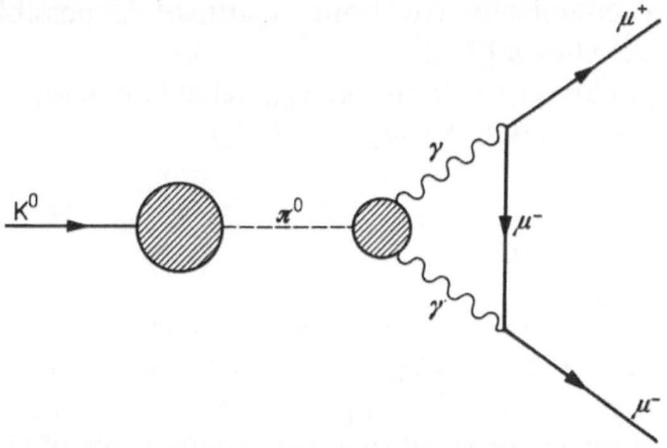

Fig. 7. Diagram for a first order weak-second order electromagnetic decay mechanism for $K_L \to \mu^+ + \mu^-$

Before going into the calculation, we mention that the decay (31) can also proceed in first order UFI with an additional second order electromagnetic interaction as shown in Fig. 7. The mechanism of Fig. 7 has been studied by *Bèg* [19], who predicted the branching ratio to be

$$\Gamma(K_L \to \mu^+ + \mu^-)/\Gamma(K_L \to \text{all}) \approx 2 \cdot 10^{-8}. \tag{32}$$

Recently, *Mohapatra et al.* [1] have reconsidered the problem and confirmed essentially *Bèg*'s value (32). The present experimental limit is [20]

$$\Gamma(K_L \to \mu^+ + \mu^-)/\Gamma(K_L \to \text{all}) < 1.6 \cdot 10^{-6}. \tag{33}$$

The matrix element M corresponding to Fig. 6a is easily written down. Treating the leptons in perturbation theory, we have:

$$M = \left(\frac{G}{\sqrt{2}}\right)^2 \cos\theta_c \sin\theta_c \int \frac{d^4 q}{(2\pi)^4} \bar{u}(p_1)\, \gamma_\mu \frac{1}{\gamma q + \gamma p_1} \gamma_\nu \cdot (1+\gamma_5)\, v(p_2)\, M_{\mu\nu}(p, q) \tag{34}$$

with

$$M_{\mu\nu} = i \int d^4x \, e^{iqx} \langle 0 | T\{V_\mu^{\pi^+}(x) + A_\mu^{\pi^+}(x), \\ V_\nu^{K^-}(0) + A_\nu^{K^-}(0)\} | K^0(\boldsymbol{p})\rangle . \tag{35}$$

Here θ_c is the Cabibbo angle, p, p_2 and p_1 are the momenta of the K, μ^- and μ^+, respectively. $V_\mu^{\pi^+}$, $V_\nu^{K^-}$ and $A_\mu^{\pi^+}$, $A_\nu^{K^-}$ are the vector- and axial current operators with their $SU(3)$-quantum numbers indicated by the upper indices ($V_\mu^{\pi^+} = V_\mu^1 - iV_\mu^2$, etc.).

In order to study the convergence of the d^4q-integration we have to investigate the high q-behaviour of $M_{\mu\nu}$. We first consider the non-covariant limit $q_0 \to \infty$ for fixed $\boldsymbol{q}$, the covariance will be restored afterwards (A completely covariant treatment is possible but more complicated, see *Olesen* [21]).

According to *Bjorken* [3], the high q_0-behaviour of $M_{\mu\nu}$ is determined by the equal time commutator ($J_\mu = V_\mu + A_\mu$):

$$\lim_{\substack{q_0 \to \infty \\ \boldsymbol{q}=0}} M_{\mu\nu} = \frac{1}{q_0} \int d^3x \langle 0 | [J_\mu^{\pi^+}(x), J_\nu^{K^-}(0)]_{x_0=0} | K^0(\boldsymbol{p})\rangle . \tag{36}$$

The commutator in (36) can be evaluated using e.g. the model of field algebra or the quark current algebra. For simplicity, in the following we shall assume a model like field algebra, which gives commuting space components of the currents, so that only components of (36) with $\mu = 4$ or $\nu = 4$ or $\mu = \nu = 4$ are nonvanishing. As we need only the matrix element of the commutator between the pseudoscalar K and the vacuum, we can drop the terms $[V_\mu, V_4]$ and $[A_\mu, A_4]$ which do not give an axial vector operator. So we obtain

$$\begin{aligned} &\int d^3x \langle 0 | [J_\mu^{\pi^+}(x), J_\nu^{K^-}(0)]_{x_0=0} | K^0(\boldsymbol{p})\rangle \\ &\quad = \int d^3x \langle 0 | \{[\delta_{\mu 4} J_4^{\pi^+}(x), J_\nu^{K^-}(0)] + [J_\mu^{\pi^+}(x), \delta_{\nu 4} J_4^{K^-}(0)] \\ &\quad\quad - \delta_{\mu 4}\delta_{\nu 4}[J_4^{\pi^+}(x), J_4^{K^-}(0)]\}_{x_0=0} | K^0\rangle \\ &\quad = 2 \langle 0 | i\,\delta_{\mu 4} A_\nu^{\bar{K}^0}(0) + i\,\delta_{\nu 4} A_\mu^{\bar{K}^0}(0) - i\,\delta_{\mu 4}\,\delta_{\nu 4} A_4^{\bar{K}^0}(0) | K^0\rangle . \end{aligned} \tag{37}$$

The matrix elements in the last line of (37) are determined by the number f_K, which is measurable in the decay $K \to \mu + \nu$:

$$\langle 0 | A_\mu^{\bar{K}^0}(0) | K^0(\boldsymbol{p})\rangle = i f_K p_\mu . \tag{38}$$

So we get

$$\lim_{\substack{q_0 \to \infty \\ \boldsymbol{q}=0}} M_{\mu\nu} = -\frac{2 f_K}{q_0} (p_\mu \delta_{\nu 4} + p_\nu \delta_{\mu 4} - p_4 \delta_{\mu 4} \delta_{\nu 4}) . \tag{39}$$

We go back to a general frame of reference by the substitution

$$\frac{i\,\delta_{\mu 4}}{q_0} \to -\frac{q_\mu}{q^2} \tag{40}$$

and obtain:

$$\lim_{q_\mu \to \infty} M_{\mu\nu} = -\frac{2if_K}{q^2}\left(p_\mu q_\nu + p_\nu q_\mu - (pq)\frac{q_\mu q_\nu}{q^2}\right). \tag{41}$$

This last step is justified from (36) only for timelike q, but we shall assume (41) to be valid for all q.

Inserting (41) into (34) we find that M is given by a divergent integral, although strong interaction effects have not been neglected in the evaluation of (36). The most divergent part of M will be called M^{Div}, it is given by:

$$M^{\mathrm{Div}} = -i\left(\frac{G}{\sqrt{2}}\right)^2 \cos\theta_c \sin\theta_c \cdot 4f_K \int \frac{d^4q}{(2\pi)^4 q^2}\,\overline{u}(p_1)\left(\gamma p - \frac{(pq)(\gamma p)}{2q^2}\right) \cdot (1+\gamma_5)\, v(p_2)\,. \tag{42}$$

In order to evaluate the integral in (42), we rotate the contour of the q_0-integration through 90^0 in the complex q_0-plane, remembering that the singularity originated from the propagator in (34). Then we introduce four-dimensional euclidian polar coordinates and cut-off the radial integration at $\varrho = \Lambda$ in order to render the integral finite:

$$\int_{-\infty}^{+\infty} \frac{d^4q}{q^2 - i\varepsilon} = i\int_0^{\infty} \frac{2\pi^2 \varrho^3\, d\varrho}{\varrho^2} = i\pi^2 \Lambda^2\,. \tag{43}$$

So we obtain for M^{Div} neglecting the term $(pq)(\gamma q)/2q^2$ in (42):

$$M^{\mathrm{Div}} = \frac{G}{\sqrt{2}} \cos\theta_c \sin\theta_c f_K \overline{u}(p_1)\gamma p(1+\gamma_5)\, v(p_2) \frac{G\Lambda^2}{4\sqrt{2}\pi^2}\,. \tag{44}$$

If we assume that the cut-off dependent term M^{Div} represents the main contribution to the whole matrix element M for the decay $K_L \to \mu^+ + \mu^-$, we can use the experimental upper limit on the decay rate $\Gamma(K_L \to \mu^+ + \mu^-)$ to put a limit on the cut-off parameter Λ. This is most easily obtained by comparing (44) to the matrix element for the well-known decay $K^+ \to \mu^+ + \nu$, which is given by

$$M_{K\to\mu+\nu} = \frac{G}{\sqrt{2}} \sin\theta_c \cdot f_K \overline{u}(p_1)\gamma p(1+\gamma_5)\, v(p_2)\,. \tag{45}$$

Taking into account the different mass of one of the final particles, one obtains for the ratio R of the decay rates:

$$\begin{aligned} R &= \frac{\Gamma(K_L \to \mu^+ + \mu^-)}{\Gamma(K^+ \to \mu^+ + \nu)} \\ &= 2\left(\frac{G\cos\theta_c \Lambda^2}{4\sqrt{2}\pi^2}\right)^2 \frac{\sqrt{1-4(m_\mu/m_K)^2}}{(1-(m_\mu/m_K)^2)^2} \approx 6\cdot 10^{-14}\left(\frac{\Lambda}{m_N}\right)^4. \end{aligned} \tag{46}$$

The experimental limit (33) can be expressed in terms of a limit on R:

$$R_{\exp} < 6\cdot 10^{-7}. \tag{47}$$

Using (46) this gives a limit on the cut-off parameter Λ:

$$\Lambda \lesssim 50\ \text{GeV}. \tag{48}$$

Thus, in order not to get into conflict with the known experimental data, we have to modify the UFI already at energies which are much lower than (22).

If, instead of (37), we assume that the current commutation relations are given by the quark model, the space components in (36) no longer commute and somewhat more complicated expressions result. The final form (41), (44), however, is the same apart from a factor 5/8 [2].

If we calculate our decay $K_L \to \mu^+ + \mu^-$ in fourth order of the IVB-hamiltonian, Fig. 6b, we also obtain the form (44) for the most divergent part of the matrix element, only this time with a factor of 1/8. The mass of the W cancels in the result for M^{Div} because not only the low q_μ-limit considered in (5)

$$\lim_{q_\mu\to 0} g^2 \frac{\delta_{\mu\nu} + q_\mu q_\nu/m_W^2}{q^2+m_W^2} = \frac{g^2\,\delta_{\mu\nu}}{m_W^2} = \left(\frac{G}{\sqrt{2}}\right)^2 \delta_{\mu\nu} \tag{49}$$

but also the high q_μ-limit

$$\lim_{q_\mu\to\infty} g^2 \frac{\delta_{\mu\nu} + q_\mu q_\nu/m_W^2}{q^2+m_W^2} = \frac{g^2 q_\mu q_\nu}{m_W^2 q^2} = \left(\frac{G}{\sqrt{2}}\right)^2 \frac{q_\mu q_\nu}{q^2} \tag{50}$$

is independent of m_W.

8. Other Estimates of the Weak Interaction Cut-Off

If we assume that the interaction hamiltonian (1) obeys the selection rule $\Delta S \leqq 1$, the virtual transition $K^0 \leftrightarrow \bar{K}^0$ (which gives rise to the observed $K_L - K_S$-mass difference) can only occur in second order UFI. The precise value of the $K_L - K_S$-mass difference is experimentally known (not only an upper limit as in the case of $K_L \to \mu^+ + \mu^-$) and it is very interesting to compare the experimental result to calculations of the second order weak transition $K^0 \leftrightarrow \bar{K}^0$. Unfortunately, because in the mass difference problem only hadrons are involved, one cannot treat part of the process in perturbation theory as we did in Fig. 6a. This time one must really calculate the matrix element of a product of four currents. This complicated problem has been discussed by several authors [1, 2, 22] in the approximation of inserting only the most simple intermediate states. In this way one finds that also in the case of the $K_L - K_S$-mass difference, the strong interactions do not make the result convergent, neither in UFI nor in IVB theory.

If one produces a finite result by the same method as we used in (43), i.e. by cutting off the integrals over the internal momenta, the experimental value of the mass difference is obtained for a very low value of the cut-off parameter Λ. The different authors agree that the value needed is

$$\Lambda = 2 \cdots 8 \text{ GeV}. \tag{51}$$

Obviously, this result is not in conflict with the estimate (48), but it indicates that the branching ratio for the second order weak decay $K_L \to \mu^+ + \mu^-$ should be much smaller than the present experimental limit (33).

We finally mention that *Glashow*, *Schnitzer* and *Weinberg* [23] some time ago estimated a similar low weak interaction cut-off value by means of a quite different method. They treated the decay $K_L \to \pi^+ + \pi^-$ by soft pion and soft kaon techniques, saturating spectral representation sum rules by single particle states. In Ref. [23], however, the cut-off appears in form of the mass of the W-meson which does not cancel from the final result.

9. Conclusion

It remains unclear what the meaning of the parameter Λ is and whether it is connected to the value of the mass of the intermediate boson. However, the necessity of using quite low values for Λ is an indication that not only at several hundred GeV, but already at energies of a few GeV an unexpected behaviour of the weak interaction should be present. This makes us hoping that already the next generation of neutrino experiments should yield new unexpected results on weak interactions.

If one follows the general opinion that the mass of a particle is somehow generated by the interactions of the particle, and if one assumes that W-mesons exist with high mass, one has to wonder which interaction may be able to produce the high W-mass. In order to explain this, *Marshak* and Coworkers have considered the hypothesis that besides the weak interaction (4) the W-mesons have strong interactions among themselves [24]. It is interesting to observe that assuming strong interactions of the W's one gets into much less difficulties than one might expect at first sight. Possibly such strong interactions which are intimately connected to the weak interactions are able to provide the cut-off of the intermediate momenta in the $K_L \to \mu^+ + \mu^-$ and $K_L - K_S$-mass difference problems which we described schematically by the parameter Λ.

References

1. *Ioffe, B. L., Shabalin, E. P.*: Yad. Fiz **6**, 828 (1967), translation: Soviet J. Nucl. Phys. **6**, 603 (1968).
2. *Mohapatra, R. N., Rao, J. S., Marshak, R. E.*: Phys. Rev. **171**, 1502 (1968).
3. *Bjorken, J. D.*: Phys. Rev. **148**, 1467 (1966).
4. *Lee, T. D., Yang, C. N.*: Phys. Rev. **108**, 1611 (1957).
5. *Derenzo, S. E., Hildebrand, R. H.*: Phys. Rev. Letters **20**, 614 (1968).
6. *Bailin, D.*: Nuovo Cimento A **40**, 822 (1965).
7. *Wu, A. C. T., Yang, C. P., Fuchel, K., Heller, S.*: Phys. Rev. Letters **12**, 57 (1964).
   Bell, J. S., Veltman, M.: Phys. Letters **5**, 94 (1963).
8. *v. Gehlen, G.*: Nuovo Cimento **30**, 859 (1963).
9. — unpublished.
10. For two recent surveys of high-energy neutrino physics see
    Perkins, D. H.: Proc. 1968 CERN School El Escorial CERN report 68–23;
    — Topical Conf. on Weak Interactions CERN 1969.
11. *Franzinetti, C.*: Topical Conf. on Weak Interactions CERN 1969.
12. *Danby, G., et al.*: Phys. Rev. Letters **9**, 36 (1962).
    Bienlein, J. K., et al.: Phys. Letters **13**, 80 (1964).
    Bernardini, G., et al.: Phys. Letters **13**, 86 (1964).
13. *Burns, R., et al.*: Phys. Rev. Letters **15**, 830 (1965).
14. *Feynman, R. P., Gell-Mann, M.*: Stanford Meeting of the Americ. Phys. Soc. 1957.
    Feinberg, G.: Phys. Rev. **110**, 1482 (1958).
15. *Pontecorvo, B.*: J. Expt. Theoret. Phys. **37**, 1751 (1959), translation: Soviet Phys. JETP **10**, 1236 (1960).
16. *Gell-Mann, M., Goldberger, M., Kroll, L., Low, F.*: Phys. Rev. **179**, 1518 (1969).
17. *Pais, A.*: Phys. Rev. Letters **9**, 117 (1962).
18. *Lee, T. D., Yang, C. N.*: Phys. Rev. **126**, 2239 (1962).
19. *Bèg, M. A. B.*: Phys. Rev. **132**, 426 (1963).
20. *Bott-Bodenhausen, M., et al.*: Phys. Letters B **24**, 194 (1967).
21. *Oleson, P.*: Phys. Rev. **172**, 1461 (1968).
22. *Low, F. E.*: Comm. Nucl. Part. Phys. **2**, 33 (1968).
23. *Glashow, S. L., Schnitzler, H. J., Weinberg, S.*: Phys. Rev. Letters **19**, 205 (1967).
24. *Pepper, S. V., Ryan, C., Okubo, S., Marshak, R. E.*: Phys. Rev. **137**, B 1259 (1965).
    Marshak, R. E., et al.: Topical Conf. on Weak Interactions CERN 1969, CERN report 69-7.

Prof. Dr. *G. von Gehlen*
Physikalisches Institut
der Universität
D-5300 Bonn, Nussallee 12

Cabibbo Angle and $SU_2 \times SU_2$ Breaking

R. GATTO

Contents

Introduction and Summary

The present notes are based on lectures given by the author at the International Institute in Theoretical Physics at Karlsruhe in July 1969. Chapters I, II and III follow very closely the lectures both in the content and in the form of the exposition. The content of Chapter IV and V was dealt with in a rather shortened form during the lectures and is here presented including more details.

Not much effort has been put in rewriting the notes, as originally prepared for the lectures, so that the presentation may in some points seem redundant in mathematical steps and relatively poor in illustrating critically the content, while imperfect in both ways in other points. Also there may be some discontinuities in the tone of the presentation, some parts being dealt with at lengthy, other parts, related to more formal developments, being presented in a rather concise form.

Part of the content of Chapter IV, and all of Chapter V, appear here for the first time. They are based on unpublished work by *G. Sartori*, *M. Tonin*, and the author, carried out during last year.

For convenience we give here a schematic outline of the content of the present notes.

We shall first develop in some detail the calculation of the quadratic divergent terms in a process involving emission and reabsorption of an intermediate *W*-boson. This is a second order process in the semiweak coupling of *W* to the currents. The model for weak interactions will be the simplest one compatible with experiment: a charged *W*-field coupled to Cabibbo currents [1]. The calculation of the quadratic divergent terms requires, in addition, the specification of the form of the $SU_3 \times SU_3$ breaking. We shall examine this problem and discuss the determination of the relevant parameters.

The second order calculation of the leading divergences suggests some speculations allowing to relate the value of the Cabibbo angle to the ratio of the parameters specifying $SU_3 \times SU_3$ breaking. An essential role is here played by an SU_2-breaking term whose presence is required to allow for a consistent scheme and which has the character of an electromagnetic tadpole. Specifically, in the original attack to the problem [2], a breaking term of the form

$$\varepsilon_0 u_0 + \varepsilon_8 u_8 + \varepsilon_3 u_3 ,$$

where the u's belong to $(3, \overline{3}) \oplus (\overline{3}, 3)$, was introduced in the hamiltonian density and the assumption was made that there should be no quadratic divergences in physical amplitudes which are not SU_3-invariant (i.e. remaining quadratic divergences only add to SU_3-invariant terms). The requirement lead us to the formula

$$\mathrm{tg}^2\theta \cong -\frac{1}{3}\left(1+\frac{\sqrt{2}\varepsilon_0}{\varepsilon_8}\right)$$

in very good agreement with experiment.

The problem that immediately appeared after this first attempt is that of dealing with higher order leading weak divergences. This requires, first of all, a detailed mathematical study [3] of such divergences for the particular class of models considered for the strong and weak interactions.

A simplest model is the free quark model (free quarks coupled to W through a Cabibbo current) for which one succeeds in establishing that, at each order, all leading divergences will add up to a unique linear superposition of mass terms (including transition masses) multiplied by a divergent coefficient depending on the order of perturbation. The problem then arises of finding out whether such a result, in itself of the greatest significance in its simplicity, is general or what is its range of validity.

Perhaps contrary to some expectations, one finds that the result is generally not valid in other models, but a class of models can be selected for which it holds either in its original form or in a slightly generalized form. This will lead us to a discussion of special limits of Lagrangian field theory which are suggestive of a composite particle picture. Proofs for the latter statements will be carried out in full rigour up to fourth order. Unfortunately the extension to higher orders (except for the case of the free quark model) is extremely cumbersome and, although the final results are almost certainly correct, the proofs lack in rigor, essentially because they involve formal manipulations with local fields which are generally unjustifiable. Assuming, provisionally, the correctness of the result, one is then in a position to verify the possibility of extending to the higher orders the speculations which lead to the determination of the Cabibbo angle.

The approach presented here is consistent in the sense that, provided a single counterterm is introduced, one has eliminated the leading divergences and the Cabibbo angle is such as to provide a "maximal" simplification of such a counterterm. One may consider such an approach as physically unsatisfactory, as it is based on the introduction of a counterterm and on its simplest choice. We shall therefore discuss an alternative approach in the last chapter including speculations leading to determinations of the Cabibbo angle. Perhaps more pleasing on their physical content, such approaches reveal themselves as still uncomplete and lacking of some essential ingredients. They seem however to suggest that one is actually close to a final solution. We hope that further work will bring new ideas and results.

I. Leading Weak Divergences at Lowest Order

1. Second Order Virtual W Process

We consider the graph

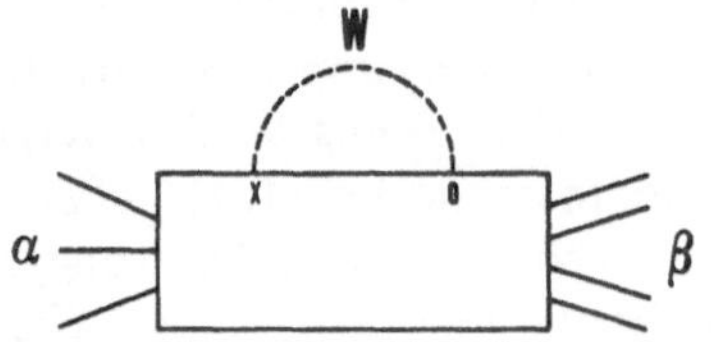

[α, β are hadronic states and, of course, $P_\alpha = P_\beta$].
The matrix element is:

$$\langle\alpha|M|\beta\rangle = -ig^2 \int \frac{d^4q}{q^2 - M^2}\left(g_{\mu\nu} - \frac{q_\mu q_\nu}{M^2}\right) \cdot \int d^4x e^{iqx} \langle\alpha| T(j_\mu(x) j_\nu^\dagger(0) + j_\mu^\dagger(x) j_\nu(0))|\beta\rangle . \tag{1.1}$$

M is the W mass and j_μ is the Cabibbo current. We assume only one charged W field coupled to a charged weak current. One should subtract disconnected diagrams such as

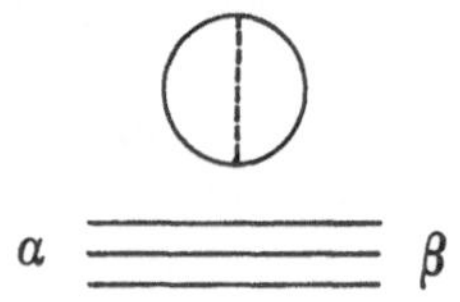

To extract the leading divergences we keep only the $q_\mu q_\nu$ term in the W propagator

$$g_{\mu\nu} - \frac{q_\mu q_\nu}{M^2} \to -\frac{q_\mu q_\nu}{M^2} .$$

One has:

$$\langle\alpha|M|\beta\rangle = \frac{ig^2}{M^2} \int \frac{d^4q}{q^2} q_\mu q_\nu \int d^4x e^{iqx} \langle\alpha| T(j_\mu(x) j_\nu^\dagger(0) + j_\mu^\dagger(x) j_\nu(0))|\beta\rangle , \tag{1.2}$$

$$\frac{g^2}{M^2} \sim G .$$

We shall now deal with the extraction of the leading divergences. Let us consider the operator:

$$iq_\mu \int d^4x e^{iqx} T(j_\mu(x) j_\nu^\dagger(0) + j_\mu^\dagger(x) j_\nu(0)) . \tag{1.3}$$

One has

$$\int d^4x \left(\frac{\partial}{\partial x_\mu} e^{iqx}\right) T(\ldots) = -\int d^4x e^{iqx} \frac{\partial}{\partial x_\mu} T(\ldots)$$
$$= -\int d^4x e^{iqx}\, \delta(x_0) \{[j_0(x), j_\nu^\dagger(0)] + [j_0^\dagger(x), j_\nu(0)]\} \qquad (1.4)$$
$$- \int d^4x e^{iqx}\, T(\mathscr{D}(x) j_\nu^\dagger(0) + \mathscr{D}^\dagger(x) j_\nu(0))\,;$$

where

$$\mathscr{D}(x) = \frac{\partial j_\mu(x)}{\partial x_\mu}. \qquad (1.5)$$

Let us assume for the equal time commutator of two currents components, the expression

$$[j_0(x), j_\nu^\dagger(0)]\, \delta(x_0) = O_\nu(x)\, \delta(x) + \delta_{\nu i} S_i \frac{\partial}{\partial x_i} \delta(x) + \cdots. \qquad (1.6)$$

For the first integral in Eq. (1.4) we can write

$$\int d^4x e^{iqx}\, \delta(x_0)\, [j_0(x), j_\nu^\dagger(0)] = \int d^4x e^{iqx}\, O_\nu(x)\, \delta(x)$$
$$+ \delta_{\nu i} S_i \int d^4x e^{iqx} \frac{\partial}{\partial x_i} \delta(x) + \cdots \qquad (1.7)$$
$$= O_\nu(0) - \delta_{\nu i} S_i i q_i + \cdots.$$

The first term does not contribute to the integral of Eq. (1.2). It would give a vanishing contribution

$$O_\nu(0) \frac{g^2}{M^2} \int \frac{d^4q}{q^2} q_\nu = 0$$

because the integral is odd in q.

The other term in Eq. (1.7) contributes

$$-i\delta_{\nu i} S_i \int \frac{d^4q}{q^2} q_\nu q_i = -i\delta_{\nu i} S_i \int \frac{d^4q}{q^2} q_i^2 \neq 0$$

in general non-vanishing.

However, if we assume that S_i is a c-number, this contribution would be eliminated by subtraction of the disconnected diagrams. When dealing with higher orders we shall actually use a more sophisticated argument to get rid of Schwinger terms. It is clear from Eq. (1.7) that Schwinger terms, if present, would add a non-covariant contribution. The final result however must be covariant. The loss of covariance arises from having defined the matrix element in terms of a T-product, which in presence of Schwinger terms is responsible for the non-covariance. By suitable modifying the T-product by substraction of contact terms one regains covariance. So in Eq. (1.4) I am left only with the last term:

$$\begin{aligned} iq_\mu \int d^4x e^{iqx} T(\ldots) &= -\int d^4x e^{iqx} T(\mathscr{D}(x) j_\nu^\dagger(0) + \mathscr{D}^\dagger(x) j_\nu(0)) \\ &= -\int d^4x e^{iqx} T(\mathscr{D}(0) j_\nu^\dagger(-x) + \mathscr{D}^\dagger(0) j_\nu(-x)) \\ &= -\int d^4x e^{-iqx} T(\mathscr{D}(0) j_\nu^\dagger(x) + \mathscr{D}^\dagger(0) j_\nu(x)) \end{aligned}$$

(we are always considering states α and β of equal fourmomentum).

We go on in our procedure by applying q_ν. We have:

$$-q_\nu \int d^4x e^{-iqx} T(\mathscr{D}(0) j_\nu^\dagger(x) + \mathscr{D}^\dagger(0) j_\nu(x))$$

(still to be integrated by applying $\int \frac{d^4q}{q^2} \cdots$, and to be multiplied by g^2/M^2).

By standard procedure

$$\begin{aligned} &= +i \int d^4x e^{-iqx} \frac{\partial}{\partial x_\nu} T(\mathscr{D}(0) j_\nu^\dagger(x) + \mathscr{D}^\dagger(0) j_\nu(x)) \\ &= i \int d^4x e^{-iqx} T(\mathscr{D}(0) \mathscr{D}^\dagger(x) + \mathscr{D}^\dagger(0) \mathscr{D}(x)) \\ &\quad + i \int d^4x e^{-iqx} \delta(x_0) \{[\mathscr{D}(0), j_0^\dagger(x)] + [\mathscr{D}^\dagger(0), j_0(x)]\}\,. \end{aligned} \tag{1.8}$$

Let us first discuss the last term in Eq. (1.8).

$$\delta(x_0)\,[\mathscr{D}(0), j_0^\dagger(x)] = O(x)\,\delta(x) + \text{Schwinger terms which are irrelevant here if } c\text{-numbers} \tag{1.9}$$

(in the models we shall consider such commutator does not have Schwinger terms at all). We have for the last term in Eq. (1.8)

$$i \int d^4x e^{-iqx} \delta(x)\,[O(x) - O^\dagger(x)] = [O(0) - O^\dagger(0)] \tag{1.10}$$

since

$$\delta(x_0)\,[\mathscr{D}^\dagger(0), j_0(x)] = -O^\dagger(x)\,\delta(x)\,.$$

The contribution of such a term is therefore

$$i[O(0) - O^\dagger(0)] \frac{g^2}{M^2} \int \frac{d^4q}{q^2}\,. \tag{1.11}$$

Upon integrating Eq. (1.9) we find

$$\int d^3x \delta(x_0)\,[\mathscr{D}(0), j_0^\dagger(x)] = [\mathscr{D}(0), Q^\dagger(x_0)]\,\delta(x_0) = O(0)\,\delta(x_0)\,,$$

where $Q(x_0) = \int d^3x j_0(x)$ and

$$(11) = i\{[\mathscr{D}(0), Q^\dagger(0)] + [\mathscr{D}^\dagger(0), Q(0)]\} \frac{g^2}{M^2} \int \frac{d^4q}{q^2}\,. \tag{1.12}$$

This turns out to be the most divergent term (quadratically divergent). We have still to consider the first term in Eq. (1.8):

$$i \int d^4x e^{-iqx} T(\mathscr{D}(0)\,\mathscr{D}^\dagger(x) + \mathscr{D}^\dagger(0)\,\mathscr{D}(x)) . \tag{1.13}$$

For this we recall in the next section an application of the Riemann-Lebesgue theorem that is due to *Bjorken* [4], and to *Johnson* and *Low* [5].

2. Bjorken-Johnson-Low Limit

Let us consider a time-ordered product $T(A(x)\,B(0))$

$$\begin{aligned} F(x) = F(\boldsymbol{x}, x_0) &= T(A(x), B(0)) \\ &= \theta(x_0)\,A(x)\,B(0) + \theta(-x_0)\,B(0)\,A(x) . \end{aligned} \tag{2.1}$$

The discontinuity from the θ functions is

$$\begin{aligned} \Delta F(\boldsymbol{x}, x_0)\,\delta(x_0) &= [F(\boldsymbol{x}, 0+) - F(\boldsymbol{x}, 0-)]\,\delta(x_0) \\ &= [A(x)\,B(0) - B(0)\,A(x)]\,\delta(x_0) = [A(x), B(0)]\,\delta(x_0) . \end{aligned} \tag{2.2}$$

Let us consider the limit for large k_0 and $\boldsymbol{k}$ fixed of

$$\int d^4x e^{-ikx}\, T(A(x), B(0)) = \int d^3x e^{-i\boldsymbol{k}\cdot\boldsymbol{x}} \int_{-\infty}^{+\infty} dx_0 e^{ik_0x_0}\, F(\boldsymbol{x}, x_0) . \tag{2.3}$$

By partial integration, formally, we obtain

$$\begin{aligned} \int_{-\infty}^{+\infty} dx_0 e^{ik_0x_0}\, F(\boldsymbol{x}, x_0) &= \frac{1}{ik_0} \int_{-\infty}^{+\infty} \left(\frac{\partial}{\partial x_0} e^{ik_0x_0}\right) F(\boldsymbol{x}, x_0)\, dx_0 \\ &= -\frac{1}{ik_0} \int_{-\infty}^{+\infty} e^{ik_0x_0}\, \dot{F}(\boldsymbol{x}, x_0)\, dx_0 + \frac{1}{ik_0}(-F(\boldsymbol{x}, 0+) + F(\boldsymbol{x}, 0-)) \\ &= -\frac{1}{ik_0} \int_{-\infty}^{+\infty} \Delta F(\boldsymbol{x}, x_0)\,\delta(x_0)\, dx_0 - \frac{1}{ik_0} \int_{-\infty}^{+\infty} e^{ik_0x_0}\, \dot{F}(\boldsymbol{x}, x_0)\, dx_0 . \end{aligned}$$

Repeating the formal process

$$\begin{aligned} \int_{-\infty}^{+\infty} dx_0 e^{ik_0x_0}\, F(\boldsymbol{x}, x_0) = &-\frac{1}{ik_0} \int_{-\infty}^{+\infty} \Delta F(x)\,\delta(x_0)\, dx_0 \\ &- \frac{1}{ik_0}\left(-\frac{1}{ik_0} \int_{-\infty}^{+\infty} \Delta\dot{F}(x)\,\delta(x_0)\, dx_0\right) + \cdots \end{aligned}$$

and upon inserting into Eq. (2.3)

$$\begin{aligned} \int d^4x e^{ikx}\, T(A(x), B(0)) = &-\frac{1}{ik_0} \int d^4x e^{ikx}\, \delta(x_0)\,[A(x), B(o)] \\ &+ \frac{1}{(ik_0)^2} \int d^4x e^{ikx}\, \delta(x_0)\,[\dot{A}(x), B(0)] + \cdots . \end{aligned}$$

3. Quadratic Divergent Terms

Going back to our case, see Eq. (1.13), we can write

$$\int d^4x e^{-iqx}\, T(\mathscr{D}(0)\,\mathscr{D}^\dagger(x)+\mathscr{D}^\dagger(0)\,\mathscr{D}(x))$$

$$\to \frac{1}{iq_0}\int d^4x e^{-iqx}\,\delta(x_0)\,\{[\mathscr{D}(0),\mathscr{D}^\dagger(x)]+[\mathscr{D}^\dagger(0),\mathscr{D}(x)]\} \tag{3.1}$$

$$+\frac{1}{(iq_0)^2}\int d^4x e^{-iqx}\,\delta(x_0)\,\{[\mathscr{D}(0),\dot{\mathscr{D}}^\dagger(x)]+[\mathscr{D}^\dagger(0),\dot{\mathscr{D}}(x)]\}\,.$$

The first term in Eq. (3.1) does not contribute upon integration, since, again assuming c-number Schwinger terms, writing

$$\delta(x_0)\,[\mathscr{D}(0),\mathscr{D}^\dagger(x)]=C(x)\,\delta(x)+\cdots$$

one has

$$\frac{1}{iq_0}\int d^4x e^{-iqx}\,\delta(x)\,[C(x)-C^\dagger(x)]=\frac{1}{iq_0}\,[C(0)-C^\dagger(0)]\,.$$

Of course, we take,

$$\int\frac{d^4q}{q}\,\frac{1}{q_0}=0$$

because the integrand is odd in q_0. The contribution from the second term in Eq. (3.1) is

$$-\int\frac{d^4q}{q^2}\,\frac{1}{q^2}\int d^4x e^{-iqx}\,\{[\mathscr{D}(0),\dot{\mathscr{D}}^\dagger(x)]+[\mathscr{D}^\dagger(0),\dot{\mathscr{D}}(x)]\}$$

and is expected to be logarithmic divergent.

Before concluding we must stress the apparent formal character of all these procedures.

4. Graphical Interpretation

Because of the complexity of the 4-order calculation to be given later, and the even greater complexity of the calculation at all orders that we shall also sketch, it is important to establish a graphical procedure which visualizes the above mathematical steps in the extraction of the most divergent terms in Eq. (1.1). The graphical interpretation is exhibited in the Figure. The graph [0] is the initial matrix element in Eq. (1.1). The operation ($q_\nu \rightarrow$) indicates taking q_ν into the integral, performing the corresponding partial integration and differentiating the T-product. At point x there is an exponential e^{iqx} under the integral corresponding to the emission of the virtual W absorbed at y, where there is an exponential factor e^{-iqy}. Of course the graphical technique here is subjected to all qualifications we have already pointed out. The point indicates there is a divergence operator, $\partial_\mu j_\mu$. The technique can be extended also in the presence of Schwinger terms.

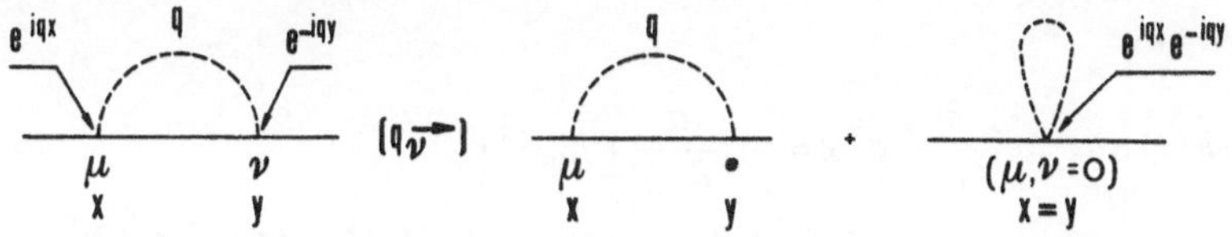

[0] [1] [2]

In [2] any dependence on q is lost (if Schwinger terms can be omitted). Therefore [2] vanishes upon integration because it is odd in q_μ.

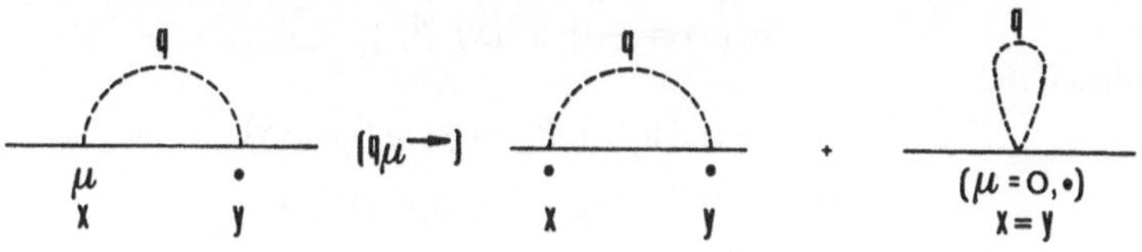

(1) (1.1) (1.2)

[1.1] is more convergent because of the oscillating factor $e^{iqx}e^{-iqy} = e^{iq(x-y)}$ and [1.2] has clearly the form $[Q, \mathscr{D}^\dagger] + [Q^\dagger, \mathscr{D}]$.

5. $SU_3 \times SU_3$ Breaking

We shall consider a hamiltonian density of the form [6]

$$\mathscr{H}(x) = \mathscr{H}_{\text{inv}}(x) + \mathscr{H}'(x) \tag{5.1}$$

where $\mathscr{H}_{\text{inv}}$ is invariant under chiral $SU_3 \times SU_3$ (of generators F_i, F_i^5) and

$$\mathscr{H}'(x) = \sum_l \varepsilon_l u_l(x) \,. \tag{5.2}$$

In Eq. (5.2) ε_l are real parameters and $u_l(x)$ are scalar densities of the $(3, \bar{3}) \oplus (\bar{3}, 3)$ representation [6]. The densities $u_l(x)$ together with the pseudoscalar densities $v_l(x)$ satisfy

$$\begin{aligned}
[F_i, u_j(x)] &= i f_{ijk} u_k(x) \,, \\
[F_i, v_j(x)] &= i f_{ijk} v_k(x) \,, \\
[F_i^5, u_j(x)] &= -i d_{ijk} v_k(x) \,, \\
[F_i^5, v_j(x)] &= i d_{ijk} u_k(x) \,.
\end{aligned} \tag{5.3}$$

Here $i=1,\dots 8$, whereas j, k run from 0 to 8. In Eq. (5.2) l is limited to $l=0, 3, 8$ because of charge and strangeness. We define

$$\mathscr{D}_i(x)=\frac{\partial j_{i\mu}(x)}{\partial x_\mu}\qquad \mathscr{D}_i^5(x)=\frac{\partial j_{i\mu}^5(x)}{\partial x_\mu}. \tag{5.4}$$

We have

$$\begin{aligned}\int \mathscr{D}_i(x)\,\mathrm{d}^3x &= \int \frac{\partial j_{i\mu}}{\partial x_\mu}\,\mathrm{d}^3x = -\frac{\partial}{\partial x_0}\int j_{i0}\,\mathrm{d}^3x \\ &= -\dot{F}_i = -i\int \mathrm{d}^3x[\mathscr{H}(x), F_i] = -i\int \mathrm{d}^3x[\mathscr{H}'(x), F_i].\end{aligned} \tag{5.5}$$

We assume that (5.5) is satisfied in the local form

$$\mathscr{D}_i(x) = -i[\mathscr{H}'(x), F_i]. \tag{5.6}$$

Similarly

$$\mathscr{D}_i^5(x) = -i[\mathscr{H}'(x), F_i^5].$$

It follows

$$\begin{aligned}\mathscr{D}_i(x) &= -i\varepsilon_j[u_j(x), F_i] = -\varepsilon_j f_{ijk}u_k(x),\\ \mathscr{D}_i^5(x) &= -i\varepsilon_j[u_j(x), F_i^5] = \varepsilon_j d_{ijk}v_k(x).\end{aligned} \tag{5.7}$$

6. Charge-divergence Commutators

The current j_μ is the Cabibbo current

$$\begin{aligned}j_\mu &= \cos\theta\,\tfrac{1}{2}(j_1+ij_2)+\sin\theta\,\tfrac{1}{2}(j_4+ij_5)\\ &\quad+\cos\theta\,\tfrac{1}{2}(j_1^5+ij_2^5)+\sin\theta\,\tfrac{1}{2}(j_4^5+ij_5^5)\end{aligned} \tag{6.1}$$

and its divergence is

$$\begin{aligned}\mathscr{D} = \frac{\partial j_\mu}{\partial x_\mu} &= \cos\theta\,\tfrac{1}{2}(\mathscr{D}_1+i\mathscr{D}_2)+\sin\theta\,\tfrac{1}{2}(\mathscr{D}_4+i\mathscr{D}_5)\\ &\quad+\cos\theta\,\tfrac{1}{2}(\mathscr{D}_1^5+i\mathscr{D}_2^5)+\sin\theta\,\tfrac{1}{2}(\mathscr{D}_4^5+i\mathscr{D}_5^5).\end{aligned} \tag{6.2}$$

The divergent expression in Eq. (1.12) is proportional to

$$[Q, \mathscr{D}^\dagger]+[Q^\dagger, \mathscr{D}]. \tag{6.3}$$

In view of later work we define the chiral generators

$$Q_i = F_i + F_i^5 \tag{6.4}$$

and the chiral divergences

$$D_i = \mathscr{D}_i + \mathscr{D}_i^5. \tag{6.5}$$

We have

$$D_i = -\varepsilon_j(f_{ijk}u_k - d_{ijk}v_k). \tag{6.6}$$

We have

$$\begin{aligned}[Q_i, D_j] &= [F_i + F_i^5, -\varepsilon_l(f_{jlk}u_k - d_{jlk}v_k)]\\ &= -\varepsilon_l f_{jlk}[F_i + F_i^5, u_k] + \varepsilon_l d_{jlk}[F_i + F_i^5, v_k]\\ &= -i\varepsilon_l f_{jlk}(f_{ikm}u_m - d_{ikm}v_m) + i\varepsilon_l d_{jlk}(f_{ikm}v_m + d_{ikm}u_m)\\ &= -i\varepsilon_l(f_{jlk}f_{ikm} - d_{jlk}d_{ikm})\,u_m + i\varepsilon_l(f_{jlk}d_{ikm} + d_{jlk}f_{ikm})\,v_m\\ &= if_{ijk}D_k + id_{ijk}E_k\,,\end{aligned} \tag{6.7}$$

where

$$E_i = \sum_l \varepsilon_l(-f_{lij}v_j + d_{lij}u_j)\,. \tag{6.8}$$

Introducing F_0 and F_0^5 and defining $Q_0 = F_0 + F_0^5$ we find that

$$E_i = -i\sqrt{\tfrac{2}{3}}[Q_0, D_i]\,. \tag{6.9}$$

Furthermore $[Q_i, E_j]$ gives again a superposition of D's and E's:

$$[Q_i, E_j] = -id_{ijk}D_k + if_{ijk}E_k\,. \tag{6.10}$$

Summarizing

$$D_i = \mathscr{D}_i + \mathscr{D}_i^5 = i[F_i + F_i^5, \mathscr{H}'] = i[Q_i, \mathscr{H}']\,, \tag{6.11}$$

$$E_i = -i\sqrt{\tfrac{2}{3}}[Q_0, D_i] = \sqrt{\tfrac{2}{3}}[Q_0[Q_i, \mathscr{H}']]\,, \tag{6.12}$$

$$[Q_i, D_j] = if_{ijk}D_k + id_{ijk}E_k\,, \tag{6.13}$$

$$[Q_i, E_j] = -id_{ijk}D_k + if_{ijk}E_k\,. \tag{6.14}$$

We now perform a Cabibbo rotation introducing

$$Q'_k = e^{2i\theta F_7}\, Q_k e^{-2i\theta F_7} = A_{kj}Q_j \tag{6.15}$$

and correspondingly

$$D'_k = A_{kj}D_j\,, \qquad E'_k = A_{kj}E_j\,. \tag{6.16}$$

Eqs. (6.13), (6.14) also hold for the set Q', D', E'. Here we are interested in an SU_2 subgroup of generators $Q'_+ = Q'_1 + iQ'_2$, $Q'_- = Q'_1 - iQ'_2$, Q'_3. We have

$$\begin{aligned}&[Q'_+, D'_+] = [Q'_-, D'_-] = 0, \quad [Q'_3, D'_3] = \frac{i}{\sqrt{3}}(\sqrt{2}E'_0 + E'_8)\,,\\ &[Q'_\pm, D'_\mp] = \pm D'_3 + \frac{i}{\sqrt{3}}(\sqrt{2}E'_0 + E'_8)\,,\\ &[Q'_\pm, D'_3] = \mp D'_\pm\,, \quad [Q'_3, D'_\pm] = \pm D'_\pm\,,\\ &\left[Q'_\pm, \frac{i}{\sqrt{3}}(\sqrt{2}E'_0 + E'_8)\right] = D'_\pm\,, \quad \left[Q'_3, \frac{i}{\sqrt{3}}(\sqrt{2}E'_0 + E'_8)\right] = D'_3\,.\end{aligned} \tag{6.17}$$

Briefly, under Q'_+, Q'_-, and Q'_3 the doublets

$$\begin{bmatrix} D'_+ \\ \\ -D'_3 + \frac{i}{\sqrt{3}}(\sqrt{2}E'_0 + E'_8) \end{bmatrix}, \quad \begin{bmatrix} D'_3 + \frac{i}{\sqrt{3}}(\sqrt{2}E'_0 + E'_8) \\ \\ D'_- \end{bmatrix} \tag{6.18}$$

transform as spinors. In the present notations our commutator Eq. (6.3), is

$$[Q'_+, D'_-] + [Q'_-, D'_+] = \frac{2i}{\sqrt{3}}(\sqrt{2}E'_0 + E'_8) \tag{6.19}$$

and from Eq. (6.18)

$$\sqrt{2}E'_0 + E'_8 = b_0 u_0 + b_3 u_3 + b_8 u_8 + b_6 u_6 + b_7 v_7 . \tag{6.20}$$

II. Chiral Symmetry Breaking

7. $SU_3 \times SU_3$ Breaking: Spectral Sum Rules

Let us go back to the model of Section 5. The symmetry breaking hamiltonian density is

$$\mathcal{H}' = \sum_l \varepsilon_l u_l . \tag{7.1}$$

Let us recall the commutators

$$\begin{aligned} [F_i, u_j] &= i f_{ijk} u_k , \\ [F_i, v_j] &= i f_{ijk} v_k , \\ [F_i^5, u_j] &= -i d_{ijk} v_k , \\ [F_i^5, v_j] &= i d_{ijk} u_k , \end{aligned} \tag{7.2}$$

and the expressions for the divergences

$$\begin{aligned} \mathcal{D}_i &= i[F_i, \mathcal{H}'] = \varepsilon_j f_{jik} u_k , \\ \mathcal{D}_i^5 &= i[F_i^5, \mathcal{H}'] = \varepsilon_j d_{jik} v_k . \end{aligned} \tag{7.3}$$

If one specializes Eq. (7.1) to

$$\mathcal{H}' = \varepsilon_0 u_0 + \varepsilon_8 u_8 \tag{7.4}$$

one has

$$\begin{aligned} \mathcal{D}_i &= \varepsilon_8 f_{8ij} u_j , \\ \mathcal{D}_i^5 &= \varepsilon_0 d_{0ij} v_j + \varepsilon_8 d_{8ij} v_j . \end{aligned} \tag{7.5}$$

Coming back to the general formalism, we introduce the spectral representations

$$\int d^4x e^{iqx} \langle T j_{i\mu}(x)\, \mathscr{D}_i(0)\rangle_0 = q_\mu \int \frac{\varrho_i(\sigma^2)\, d\sigma^2}{q^2-\sigma^2}, \tag{7.6}$$

$$i \int d^4x e^{iqx} \langle T \mathscr{D}_i(x)\, \mathscr{D}_i(0)\rangle_0 = \int \frac{\tau_i(\sigma^2) d\sigma^2}{q^2-\sigma^2} \tag{7.7}$$

(no sum on i).
We have:

$$\begin{aligned} iq_\mu \int d^4x e^{iqx} \langle T j_{i\mu}(x)\, \mathscr{D}_i(0)\rangle_0 &= - \int d^4x e^{iqx} \frac{\partial}{\partial x\mu} \langle T j_{i\mu}(x)\, \mathscr{D}_i(0)\rangle_0 \\ &= - \int d^4x e^{iqx} \langle T \mathscr{D}_i(x) \mathscr{D}_i(0)\rangle_0 - \int d^4x e^{iqx} \delta(x_0) \langle [j_{i0}(x), \mathscr{D}_i(0)]\rangle_0 \\ &= - \int d^4x e^{iqx} \langle T \mathscr{D}_i(x) \mathscr{D}_i(0)\rangle_0 - \varepsilon_j f_{jik} \int d^4x e^{iqx} \delta(x_0) \langle [j_{i0}(x), u_k(0)]\rangle_0 . \end{aligned} \tag{7.8}$$

We assume that Eqs. (7.2) also hold when one substitutes for the generators F_i and F_i^5 the corresponding local current $j_i(x)$ and $j_i^5(x)$. We thus have

$$\begin{aligned} iq_\mu \int d^4x e^{iqx} \langle T j_{i\mu}(x)\, \mathscr{D}_i(0)\rangle_0 \\ = - \int d^4x e^{iqx} \langle T \mathscr{D}_i(x)\, \mathscr{D}_i(0)\rangle_0 - i\varepsilon_j f_{jik} f_{ikl} \langle u_l\rangle_0 . \end{aligned} \tag{7.9}$$

Inserting the spectral representations of Eqs. (7.6) and (7.7) we find

$$iq^2 \int \frac{\varrho_i(\sigma^2)\, d\sigma^2}{q^2-\sigma^2} = i \int \frac{\tau_i(\sigma^2)\, d\sigma^2}{q^2-\sigma^2} - i\varepsilon_j f_{jik} f_{ikl} \langle u_l\rangle_0$$

or

$$\int \frac{d\sigma^2}{q^2-\sigma^2} (q^2 \varrho_i(\sigma^2) - \tau_i(\sigma^2)) = - \varepsilon_j f_{jik} f_{ikl} \langle u_l\rangle_0 . \tag{7.10}$$

Taking the imaginary part of Eq. (7.10) we find

$$q^2 \varrho_i(\sigma^2) - \tau_i(\sigma^2) = (q^2 - \sigma^2)\, \varrho_i(\sigma^2)$$

or

$$\tau_i(\sigma^2) = \sigma^2 \varrho_i(\sigma^2) \tag{7.11}$$

and Eq. (7.10) becomes

$$S_i \equiv \int d\sigma^2 \varrho_i(\sigma^2) = - \varepsilon_j f_{jik} f_{ikl} \langle u_l\rangle_0 . \tag{7.12}$$

Similarly, for the axial case, we define

$$\int d^4x e^{iqx} \langle T j_{i\mu}^5(x)\, \mathscr{D}_i^5(0)\rangle_0 = q_\mu \int \frac{\varrho_i^5(\sigma^2)\, d\sigma^2}{q^2-\sigma^2}, \tag{7.13}$$

$$i \int d^4x e^{iqx} \langle T \mathscr{D}_i^5(x) \mathscr{D}_i^5(0)\rangle_0 = \int \frac{\tau_i^5(\sigma^2)\, d\sigma^2}{q^2-\sigma^2} \tag{7.14}$$

(again no sum on i).

We have, for the axial-vector case:

$$iq_\mu \int d^4x e^{iqx} \langle T j_{i\ \mu}^5(x) \mathscr{D}_i^5(0)\rangle_0$$
$$= -\int d^4x e^{iqx} \langle T \mathscr{D}_i^5(x) \mathscr{D}_i^5(0)\rangle - \int d^4x e^{iqx} \delta(x_0) \langle [j_{io}^5(x), \mathscr{D}_i^5(0)]\rangle_0$$

and again, inserting the local commutators.

$$iq_\mu \int d^4x e^{iqx} \langle T j_{i\mu}^5(x) \mathscr{D}_i^5(0)\rangle_0$$
$$= -\int d^4x e^{iqx} \langle T \mathscr{D}_i^5(x) \mathscr{D}_i^5(0)\rangle_0 - i\varepsilon_j d_{jik} d_{ikl} \langle u_l\rangle_0 \tag{7.15}$$

Inserting Eqs. (7.13) and (7.14) into Eq. (7.15) we find

$$iq^2 \int \frac{\varrho_i^5(\sigma^2)\, d\sigma^2}{q^2-\sigma^2} = i \int \frac{\tau_i^5(\sigma^2)\, d\sigma^2}{q^2-\sigma^2} - i\varepsilon_j d_{jik} d_{ikl} \langle u_l\rangle_0 \tag{7.16}$$

Thus

$$\tau_i^5 = \sigma^2 \varrho_i^5 \tag{7.17}$$

and

$$S_i^5 \equiv \int \varrho_i^5(\sigma^2)\, d\sigma^2 = -\varepsilon_j d_{jik} d_{ikl} \langle u_l\rangle_0 \tag{7.18}$$

In conclusion we note that in the whole argument there was no specification of the parameters ε_j. Summarizing:

For

$$\mathscr{H}' = \Sigma_l \varepsilon_l u_l \tag{7.1}$$

one has (no sum on i)

$$S_i \equiv \int d\sigma^2 \varrho_i(\sigma^2) = -\varepsilon_{ik}^{(o)} \lambda_{ik}^{(o)} \tag{7.12}$$

$$S_i^5 \equiv \int d\sigma^2 \varrho_i^5(\sigma^2) = -\varepsilon_{ik}^{(e)} \lambda_{ik}^{(e)} \tag{7.18}$$

where the spectral functions are defined from

$$i \int d^4x e^{iqx} \langle T \mathscr{D}_i(x) \mathscr{D}_i(0)\rangle_0 = \int \frac{\sigma^2 \varrho_i(\sigma^2)\, d\sigma^2}{q^2-\sigma^2} \tag{7.19}$$

$$i \int d^4x e^{iqx} \langle T \mathscr{D}_i^5(x) \mathscr{D}_i^5(0)\rangle_0 = \int \frac{\sigma^2 \varrho_i^5(\sigma^2)\, d\sigma^2}{q^2-\sigma^2} \tag{7.20}$$

and we have introduced the combinations

$$x_{ik}^{(o)} = f_{ikl} x_l\,, \quad x_{ik}^{(e)} = d_{ikl} x_l\,, \tag{7.21}$$

where the superscripts o and e stand for "odd" and "even" respectively, corresponding to the appearance of the SU_3 coefficients f_{ijk} and d_{ijk} respectively.

8. $SU_3 \times SU_3$ Breaking into SU_2

In this case

$$\mathscr{H}' = \varepsilon_0 u_0 + \varepsilon_8 u_8 . \tag{8.1}$$

We have to specialize Eqs. (7.12) and (7.18). Also $\langle u_l \rangle = 0$ unless $l = 0{,}8$. We have

$$S_i = -\varepsilon_{ik}^{(o)} \lambda_{ik}^{(o)}, \qquad S_i^5 = -\varepsilon_{ik}^{(e)} \lambda_{ik}^{(e)}, \tag{8.2}$$

$$\varepsilon_{ik}^{(o)} = f_{ik8}\varepsilon_8, \qquad \lambda_{ik}^{(o)} = f_{ik8}\langle u_8 \rangle_0, \tag{8.3}$$

$$\varepsilon_{ik}^{(e)} = d_{ik0}\varepsilon_0 + d_{ik8}\varepsilon_8, \qquad \lambda_{ik}^{(e)} = d_{ik0}\langle u_0 \rangle_0 + d_{ik8}\langle u_8 \rangle_0 . \tag{8.4}$$

For instance

$$S_1^5 = -\varepsilon_{1k}^{(e)} \lambda_{1k}^{(e)} = -\left(\sqrt{\frac{2}{3}}\,\varepsilon_0 + \frac{1}{\sqrt{3}}\,\varepsilon_8\right)\left(\sqrt{\frac{2}{3}}\,\langle u_0 \rangle_0 + \frac{1}{\sqrt{3}}\,\langle u_8 \rangle_0\right)$$

etc.

We shall introduce the abbreviation: $\lambda_i = \langle u_i \rangle_0$. We consider the independent set of equations

$$i = 1, 2, 3, \qquad S_i^5 \equiv S_\pi = -\tfrac{1}{3}\,(\sqrt{2}\,\varepsilon_0 + \varepsilon_8)(\sqrt{2}\,\lambda_0 + \lambda_8), \tag{8.5}$$

$$i = 4, 5, 6, 7, \qquad S_i^5 \equiv S_K = -\tfrac{1}{3}(\tfrac{1}{2}\,\varepsilon_8 - \sqrt{2}\,\varepsilon_0)(\tfrac{1}{2}\,\lambda_8 - \sqrt{2}\,\lambda_0), \tag{8.6}$$

$$i = 4, 5, 6, 7, \qquad S_i \equiv S = -\tfrac{3}{4}\,\varepsilon_8 \lambda_8 . \tag{8.7}$$

The parameter $\sqrt{2}\,\varepsilon_0 + \varepsilon_8$ is zero when the hamiltonian is invariant under $SU_2 \times SU_2$ chiral. In fact for $i = 1, 2, 3$

$$\mathscr{D}_i^5 = \varepsilon_j d_{jik} v_k = (\varepsilon_0 d_{0ik} + \varepsilon_8 d_{8ik})\, v_k$$
$$= \left(\sqrt{\frac{2}{3}}\,\varepsilon_0 + \frac{1}{\sqrt{3}}\,\varepsilon_8\right)\delta_{ik} v_k = \left(\sqrt{\frac{2}{3}}\,\varepsilon_0 + \frac{1}{\sqrt{3}}\,\varepsilon_8\right) v_i . \tag{8.8}$$

So that, if there is no breaking at all, or only spontaneous breaking: $\sqrt{2}\,\varepsilon_0 + \varepsilon_8 = 0$. The operators then satisfy $SU_2 \times SU_2$. It is thus convenient to introduce a parameter

$$\varrho = -\left(1 + \frac{\sqrt{2}\,\varepsilon_0}{\varepsilon_8}\right) \tag{8.9}$$

such that $\varrho = 0$ when $\sqrt{2}\,\varepsilon_0 + \varepsilon_8 = 0$, i.e. when one has exact $SU_2 \times SU_2$ in the Lagrangian. Note that exact SU_3, i.e. $\varepsilon_8 = 0$, implies $\varrho \to \infty$. From Eqs. (8.5), (8.6), and (8.7) one obtains

$$\varrho_\pm = \frac{3}{4}\,\frac{S_K - S_\pi - S \pm \sqrt{\Delta}}{S}, \tag{8.10}$$

$$\Delta = S_K^2 + S_\pi^2 + S^2 - 2S_\pi S_K - 2S_K S - 2S_\pi S . \tag{8.10'}$$

We have now to choose the right sign in Eq. (8.10). We must require that when $S_\pi \to 0$ the axial vector divergence $\mathscr{D}_i^5$ for $i=1,2,3$ vanishes. This is in fact the content of the operator formulation of PCAC. Therefore when $S_\pi \to 0$ we must have

$$\sqrt{\frac{2}{3}}\,\varepsilon_0 + \frac{1}{\sqrt{3}}\,\varepsilon_8 \to 0 \qquad (\text{for } S_\pi \to 0)$$

or equivalently

$$\varrho \to 0 \qquad (\text{for } S_\pi \to 0)\,.$$

The two solutions behave as follows for $S_\pi \to 0$

$$\varrho_+ \to 2\,\frac{S_K - S}{S}\,, \qquad \varrho_- \to 0\,.$$

Therefore the correct choice is ϱ_-. This completes the derivation. It is also interesting to consider the limit $S \to 0$. The two solution behave as follows

$$\varrho_+ \to \frac{3}{2}\,\frac{S_K - S_\pi}{S}\,,$$

$$\varrho_- \to \frac{3}{2}\,\frac{S_\pi}{S_K - S_\pi}\,.$$

Under approximate saturation one has $S_K \sim F_K^2 \mu_K^2$, $S_\pi \sim F_\pi^2 \mu_\pi^2$ that is $S_\pi \ll S_K$. For small S, as we expect from the approximate validity of SU_3 in the mass spectra, ϱ_+ is very large whereas ϱ_- is very small. This means that, if ϱ_+ is the correct choice, one has $\varepsilon_8 \approx 0$ (see Eq. (8.9)), that is a very approximate (and predominantly spontaneously broken) SU_3 invariance in the Lagrangian. If ϱ_- is the right solution the Lagrangian is close to $SU_2 \times SU_2$. Our argument on the Cabibbo angle will independently lead as we shall see, to the choice of ϱ_-. Since when $\varrho \to 0$, $\varepsilon_8 \to \sqrt{2}\,\varepsilon_0$, and is therefore not negligible (with respect to ε_0) it follows that the limit $S \to 0$ is realized (see Eq. (8.7)) through the smallness of $\lambda_8 = \langle u_8 \rangle_0$. For the correct solution, ϱ_-, we have for S_π small with respect to S_K

$$\varrho \cong \frac{3}{2}\,\frac{S_\pi}{S_K - S}\,. \tag{8.11}$$

Approximately

$$\varrho \cong \frac{3}{2}\,\frac{S_\pi}{S_K}\left(1 + \frac{S}{S_K} + \cdots\right). \tag{8.12}$$

Under approximate saturation

$$\varrho \cong \frac{3}{2}\,\frac{F_\pi^2 \mu_\pi^2}{F_K^2 \mu_K^2}\left(1 + \frac{S}{F_K^2 \mu_K^2}\right). \tag{8.13}$$

Here F_π and F_K are the usual π and K decay constants. A numerical estimate can be obtained, if following *Glashow-Weinberg* [7], one saturates S with a scalar state x

$$S \cong F_x^2 \mu_x^2 . \tag{8.14}$$

These authors give:

$$\begin{aligned} F_K^2/F_\pi^2 &= 1.17 , \\ F_x^2/F_\pi^2 &= 0.34 , \\ \mu_x &\lesssim 670 \text{ MeV} . \end{aligned} \tag{8.15}$$

One then obtains

$$\varrho = 0.15 . \tag{8.16}$$

Gell-Mann, Oakes and *Renner* [7] have given a derivation of the symmetry breaking along essentially similar lines. They calculate the parameter $c = \varepsilon_8/\varepsilon_0$.

One has

$$c = -\sqrt{2}\frac{1}{1+\varrho} . \tag{8.17}$$

For small ϱ

$$\begin{aligned} c &\cong -\sqrt{2}(1-\varrho) , \\ \varrho &\cong 1 + \frac{1}{\sqrt{2}} c . \end{aligned} \tag{8.18}$$

With our expression, Eq. (8.13), we would get

$$c \cong -\sqrt{2}\left[1 - \frac{3}{2}\frac{S_\pi}{S_K} - \frac{3}{2}\frac{S_\pi S}{S_K^2}\right] \tag{8.19}$$

and in the pole approximation

$$c \cong -\sqrt{2}\left[1 - \frac{3}{2}\frac{F_\pi^2}{F_K^2}\frac{\mu_\pi^2}{\mu_K^2} - \frac{3}{2}\frac{F_\pi^2 F_x^2}{F_K^2}\frac{\mu_\pi^2 \mu_x^2}{\mu_K^4}\right] \tag{8.20}$$

The expression by *Gell-Mann-Oakes-Renner* (GMOR) is

$$c \cong -\sqrt{2}\left(1 - \frac{3}{2}\frac{\mu_\pi^2}{\mu_K^2}\right) . \tag{8.21}$$

It is obtained from Eq. (8.20) if one puts $F_\pi^2/F_K^2 = 1$ and neglects $\mu_x^2 F_x^2$.

Numerically one has

(GMOR) $\quad c = -1.25 \quad (\varrho = 0.115)$ (8.22)

whereas, with the values in Eq. (8.15), we obtained

(GST) $\quad c = -\sqrt{2}(1 - 0.1 - 0.05) = -1.2 . \quad (\varrho = 0.15)$ (8.23)

The decomposition of the sum in Eq. (23) corresponds to the sum in Eq. (20).

The derivation in GMOR goes as follows. One has (P_i is a *ps* state), using PCAC (they use PCAC rather than our pole approximation)

$$\langle P_i(p)|\mathscr{H}'|P_i(p')\rangle = \int e^{-ipx} d^4x(\Box_x^2 + \mu_i^2)\, \langle 0|T(\phi_i(x)\mathscr{H}')|P_i(p')\rangle$$

$$\lim_{p\to 0}\langle P_i(p)|\mathscr{H}'|P_i(p')\rangle = \mu_i^2 \int e^{-ipx} d^4x\, \langle 0|T(\phi_i(x)\mathscr{H}')|P_i(p')\rangle .$$

Using

$$\mathscr{D}_i^5 = \partial_\mu j_{i\mu}^5 = F_i \mu_i^2 \phi_i$$

one has

$$\lim_{p\to 0}\langle P_i(p)|\mathscr{H}'|P_i(p')\rangle = \frac{1}{F_i}\int e^{-ipx} d^4x\, \langle 0|T(\partial_\mu j_{i\mu}^5(x)\mathscr{H}')|P_i(p')\rangle$$

By partial integration

$$\lim_{p\to 0}\langle P_i(p)|\mathscr{H}'|P_i(p')\rangle = -\frac{i}{F_i}\int d^4x\,\delta(x_0)\langle 0|[j_{i0}^5(x), \mathscr{H}']|P_i(p')\rangle$$

$$-\frac{i}{F_i}\langle 0|[F_i^5, \mathscr{H}']|P_i(p')\rangle = -\frac{1}{F_i}\langle 0|\mathscr{D}_i^5|P_i(p')\rangle = -\mu_i^2 .$$

At this point one inserts an SU_3 assumption

$$-\langle P_i(p)|u_j|P_k(p')\rangle = \beta d_{ijk} + \alpha\delta_{j0}\delta_{ik}$$

always in the low-momentum limit. β and α can be calculated from the mass formula

$$-\langle P_i(p)|u_0 + cu_8|P_i(p)\rangle = \mu_i^2 = \langle \mu^2\rangle_{\text{av}} + d^{ii8}\Delta m^2$$

and one finds

$$\beta = \frac{\Delta m^2}{c},$$

$$\alpha = \langle m^2\rangle - \sqrt{\frac{2}{3}}\,\frac{\Delta m^2}{c}.$$

For $\langle P_i(p)|u_j|P_k(p')\rangle$, always in the soft pion limit, one has

$$-\langle P_i(p)|u_j|P_k(p')\rangle = +i\frac{1}{F_i}\langle 0|[F_i^5, u_j]|P_k(p')\rangle = \frac{1}{F_i} d_{ijk}\langle 0|v_k|P_k(p')\rangle$$

and also

$$= -i\frac{1}{F_k}\langle 0|[F_k^5, u_j]|0\rangle = -\frac{1}{F_k} d_{kjl}\langle P_i|v_l|0\rangle .$$

In conclusion

$$-\langle P_i|u_j|P_k\rangle = \beta d_{ijk} + \alpha\delta_{j0}\delta_{ik} = \frac{1}{F_i} d_{ijk}\langle 0|v_k|P_k\rangle$$
$$= -\frac{1}{F_k} d_{kjl}\langle P_i|v_l|0\rangle . \tag{8.24}$$

Furthermore

$$-\langle P_i|\mathscr{H}'|P_i\rangle = \mu_i^2 = \langle\mu^2\rangle_{\text{av}} + d^{ii8}\Delta m^2 . \tag{8.25}$$

Solving Eq. (8.24) GMOR obtain:

$$\begin{aligned} F_\pi &= F_K = F_\eta = F , \\ \langle 0|v_i|P_i\rangle &= -\frac{\beta}{2F} , \\ \langle 0|v_0|\eta\rangle &= 0 , \\ \alpha &= 0 , \\ c &= \sqrt{\frac{3}{2}}\,\frac{\langle m^2\rangle}{\Delta m^2} = -1.25 . \end{aligned} \tag{8.26}$$

Explicitly

$$c = -\sqrt{2}\left(1 - \frac{3}{2}\,\frac{\mu_\pi^2}{\mu_K^2}\right)$$

also it follows

$$\lambda_8 = \langle u_8\rangle_0 \cong 0 \tag{8.27}$$

(whereas $\langle u_0\rangle_0 = \mu_K^2 + \frac{1}{2}\mu_\pi^2$).

9. Addition of an Isospin-breaking Term

We consider a more general form for the symmetry-breaking hamiltonian density, namely

$$\mathscr{H}' = \varepsilon_0 u_0 + \varepsilon_8 u_8 + \varepsilon_3 u_3 . \tag{9.1}$$

This breaking can be taken as a model of a breaking hamiltonian which contains some of the electromagnetic contributions. Taking the sum rules in Eq. (7.18) for K^+, K^0, and π one has

$$\begin{aligned} S_{K^+} &= -\frac{1}{3}\left(\frac{1}{2}\varepsilon_8 - \sqrt{2}\varepsilon_0 + \frac{\sqrt{3}}{2}\varepsilon_3\right)\left(\frac{1}{2}\lambda_8 - \sqrt{2}\lambda_0 - \frac{\sqrt{3}}{2}\lambda_3\right), \\ S_{K^0} &= -\frac{1}{3}\left(\frac{1}{2}\varepsilon_8 - \sqrt{2}\varepsilon_0 + \frac{\sqrt{3}}{2}\varepsilon_3\right)\left(\frac{1}{2}\lambda_8 - \sqrt{2}\lambda_0 + \frac{\sqrt{3}}{2}\lambda_3\right), \\ S_\pi &= -\frac{1}{3}(\sqrt{2}\varepsilon_0 + \varepsilon_8)(\sqrt{2}\lambda_0 + \lambda_8) . \end{aligned} \tag{9.2}$$

Solving in the approximation $\lambda_8 \cong 0$, $\lambda_3 \cong 0$ we obtain

$$\frac{\sqrt{3}\,\varepsilon_3}{\sqrt{2}\,\varepsilon_0+\varepsilon_8}=\frac{S_{K^+}-S_{K^0}}{S_\pi}. \tag{9.3}$$

Saturating with *ps* poles we find

$$\frac{\sqrt{3}\,\varepsilon_3}{\sqrt{2}\,\varepsilon_0+\varepsilon_8}=\frac{F_{K^+}^2\mu_{K^+}^2-F_{K^0}^2\mu_{K^0}^2}{F_\pi^2\mu_\pi^2} \tag{9.4}$$

$$=\frac{\mu_{K^+}^2-\mu_{K^0}^2}{\mu_\pi^2}+\frac{\delta F_+\mu_{K^+}^2-\delta F_0^2\mu_{K^0}^2}{F_\pi^2\mu_\pi^2} \tag{9.5}$$

where we have written for convenience

$$F_{K^+}^2=F_\pi^2+\delta F_+^2\,, \tag{9.6}$$

$$F_{K^0}^2=F_\pi^2+\delta F_0\,. \tag{9.6'}$$

Introducing the parameter ϱ we have

$$\sqrt{\frac{3}{2}}\,\frac{\varepsilon_3}{\varepsilon_0}=\varrho\left[\frac{\mu_{K^+}^2-\mu_{K^0}^2}{\mu_\pi^2}+\frac{\mu_K^2}{\mu_\pi^2}\,\frac{\delta F_+^2-\delta F_0^2}{F_\pi^2}\right] \tag{9.7}$$

$$=-\varrho\left[\frac{\mu_{K^0}^2-\mu_{K^+}^2}{\mu_\pi^2}-62\,\frac{\delta F_+^2-\delta F_0^2}{F_\pi^2}\right]. \tag{9.8}$$

Numerically:

$$\sqrt{\frac{3}{2}}\,\frac{\varepsilon_3}{\varepsilon_0}=-3.15\times 10^{-2}\left(1-62\,\frac{\delta F_+^2-\delta F_0^2}{F_\pi^2}\right)\qquad(\varrho=0.15)\,,$$

$$\sqrt{\frac{3}{2}}\,\frac{\varepsilon_3}{\varepsilon_0}=-2.4\times 10^{-2}\left(1-62\,\frac{\delta F_+^2-\delta F_0^2}{F_\pi^2}\right)\qquad(\varrho=0.115,\ \text{GMOR})\,,$$

One may be tempted to neglect the term

$$62\,\frac{\delta F_+^2-\delta F_0^2}{F_\pi^2}.$$

However, it is quite possible for such a term to be of the order of unity. For instance if $[\delta F_+^2-\delta F_0^2]\cong\alpha^2F_K^2$ one has for such a term a value ~ 0.5. Direct measurement of δF_+^2 and δF_0^2 are of course difficult, first of all they must be properly identified in the radiative corrections, Calculation of $\sqrt{\frac{3}{2}}\,\frac{\varepsilon_3}{\varepsilon_0}$ based only on the first term in the formula (9.7) are to be considered as guesses, which may eventually turn out to be right but one does not know how much they can be trusted.

10. Description in Terms of Fictitious Quark Masses

Instead of ε_0, ε_3, ε_8, we can introduce fictitious quark masses. One has only to think of the breaking as realized in terms of quark mass terms

$$-\mathscr{H}' = m_p \bar{p} p + m_n \bar{n} n + m_\lambda \bar{\lambda} \lambda . \tag{10.1}$$

The correspondence is

$$\begin{cases} -m_p = \sqrt{2}\varepsilon_0 + \varepsilon_8 + \sqrt{3}\varepsilon_3 \\ -m_n = \sqrt{2}\varepsilon_0 + \varepsilon_8 - \sqrt{3}\varepsilon_8 \\ -m_\lambda = \sqrt{2}\varepsilon_0 - 2\varepsilon_8 . \end{cases} \tag{10.2}$$

Normalizing according to $\varepsilon_0 = -1$ we have

$$\begin{aligned} m_p &= \sqrt{2}\left(\varrho - \sqrt{\frac{3}{2}}\varepsilon_3\right), \\ m_n &= \sqrt{2}\left(\varrho + \sqrt{\frac{3}{2}}\varepsilon_3\right), \\ m_\lambda &= \sqrt{2}\,(3 - 2\varrho) . \end{aligned} \tag{10.3}$$

III. Leading Weak Divergences. Free Quark Model

11. The Leading Weak Divergences (Second Order)

As we already know at second order the leading divergences are proportional to

$$\begin{aligned} [Q, \mathscr{D}^\dagger] + [Q^\dagger, \mathscr{D}] &= \frac{2i}{\sqrt{3}} (\sqrt{2} E'_0 + E'_8) \\ &= b_0 u_0 + b_3 u_3 + b_8 u_8 + b_6 u_6 + b_7 v_7 . \end{aligned} \tag{11.1}$$

We have assumed that the breaking is entirely in the $(3, \bar{3}) \oplus (\bar{3}, 3)$ representation.

The coefficients b are immediately determined from Eqs. (6.19) and (6.8). Now u_6 and v_7 are proportional to current divergences (of $j_{\mu 7}$ and $j^5_{\mu 7}$ respectively) so they will not contribute [8] to our matrix element. As far as b_0 is concerned, it gives a mass shift equal for all members of an SU_3 multiplet. We know that, in the solutions of the field equations, SU_3 (and not $SU_3 \times SU_3$) is the relevant symmetry. Therefore we are not interested in b_0, as it does not disturb such a symmetry. We are left with

$$b_3 = \frac{1}{\sqrt{3}}(\sqrt{2}\varepsilon_0 + \varepsilon_8)\sin^2\theta + \varepsilon_3(\cos^2\theta + 1),$$
$$b_8 = \frac{1}{3}(\sqrt{2}\varepsilon_0 + \varepsilon_8)(2 - 3\sin^2\theta) + 2\varepsilon_8\sin^2\theta + \frac{1}{\sqrt{3}}\varepsilon_3\sin^2\theta. \tag{11.2}$$

Requiring $b_3 = 0$ and $b_8 = 0$ gives

$$\mathrm{tg}^2\theta \cong \frac{\varrho}{3}\left(1 - \frac{\varrho}{6}\right), \tag{11.3}$$

$$\frac{\sqrt{3}\varepsilon_3}{\varepsilon_8} \cong \frac{1}{6}\varrho^2. \tag{11.4}$$

Numerically with $\varrho = 0.115$ (GMOR) one finds

$$\mathrm{tg}\theta = 0.19. \tag{11.5}$$

With our $\varrho = 0.15$

$$\mathrm{tg}\theta = 0.22. \tag{11.5'}$$

As far as Eq. (11.4) is concerned, one finds

$$\frac{\sqrt{3}\varepsilon_3}{\varepsilon_8} \cong 0.22 \times 10^{-2} \qquad (\varrho = 0.115), \tag{11.6}$$

$$= 0.38 \times 10^{-2} \qquad (\varrho = 0.15). \tag{11.6'}$$

These values are of order of magnitude comparable to $\alpha = 0.72 \times 10^{-2}$. We have assumed that the breaking belongs entirely to $(3, \bar{3}) \oplus (\bar{3}, 3)$ A more complicated form of breaking would bring additional terms in the leading divergence and complicate the whole picture [9].

In a model with free quarks one agrees as follows. The leading weak contributions to quark masses are easily calculated from the diagram for p, n, and λ. The diagram is of the kind

One has

$$\delta m_p \propto m_p \cos^2\theta + m_p \sin^2\theta = m_p, \tag{11.7}$$

$$\delta m_n \propto m_n \cos^2\theta, \tag{11.8}$$

$$\delta m_\lambda \propto m_\lambda \sin^2\theta. \tag{11.9}$$

Requiring

$$\delta m_n - \delta m_\lambda = 0 \quad \text{gives} \quad m_n \cos^2\theta = m_\lambda \sin^2\theta$$

or

$$\mathrm{tg}^2\theta = \frac{m_n}{m_\lambda}. \tag{11.10}$$

Requiring

$$\delta m_\lambda - \frac{1}{2}(\delta m_p + \delta m_n) = 0$$

gives

$$\frac{1}{m_p} = \frac{1}{m_n} + \frac{1}{m_\lambda} \tag{11.11}$$

and these equations through (10.2) are clearly identical to those above.

12. Leading Divergences at Higher Weak Orders

The interpretation of the results of the preceding sections depends of course on that viewpoint one wants to assume. Suppose for instance one takes the viewpoint that the theory is finite because of a damping mechanism at some energy much below the unitarity limit (this could happen in the model by *T. D. Lee* and *G. C. Wick* [10], for instance). The main contribution would be expected to come from the second order terms. If instead the cutoff is around the unitarity limit one has to take higher orders into account in an essential way. We first need mathematical results about the higher orders. We shall find that: In the free quark model (quark interacting with W only) all leading divergences can be accounted for by a unique counterterm which is proportional to

$$[Q, \mathscr{D}^\dagger] + [Q^\dagger, \mathscr{D}] = \sqrt{2}E_0' + E_8' = (b_0 u_0 + b_3 u_3 + b_8 u_8 + b_6 u_6 + b_7 v_7)$$

that is to the leading weak divergent term at second order. The coefficients b_i are thus known functions of the ε's and of θ. A formal derivation suggests that the same result holds in a model based on the commutators of ECFA (extended compound field algebra) that we shall describe soon after. It is then consistent to again impose the condition that b_3 and b_8 vanish in the expression for the counterterm at all orders and one finds again the second order result. We do not however regard this theory as a final and satisfactory one. In fact the origin of the counterterm should be related in our opinion to some interaction not included in the treatment. Introducing a counterterm is merely a statement of ignorance.

In the frame of the above results, it is important to note that non-leptonic decays are free of leading divergences even without introducing the counterterm. In fact the counterterms contributing to non-leptonic decays are u_6 and v_7. But both u_6 and v_7 are current divergences (from the Eqs. (5.7)). Their contributions between states of equal four-momentum (or for what matters, to any physical process) is therefore vanishing [8]. In fact if $\mathscr{D}(x)=\partial_\mu j_\mu(x)$

$$\begin{aligned}\langle\alpha|\mathscr{D}(x)|\beta\rangle &= \partial_\mu\langle\alpha|j_\mu(x)|\beta\rangle \\ &= \partial_\mu(e^{i(p_\alpha-p_\beta)x}\langle\alpha|j_\mu(0)|\beta\rangle) = i(p_\alpha-p_\beta)_\mu\langle\alpha|j_\mu(x)|\beta\rangle\,.\end{aligned} \tag{12.1}$$

The argument is more involved when dealing with time-ordered products containing an arbitrary (but, of course, odd) number of factors u_6 and/or v_7. In this case one considers a particular u_6 or v_7, substitutes for it the corresponding current divergence, and performs a partial integration in the usual way. One can see that no pole terms come in this procedure. In the resulting equal time commutator one inserts the current algebra expression. By repeated use of the procedure one finally arrives at a single u_6 or v_7, which does not contribute, by the argument given before. It is also important to discuss the dependence on the form of the symmetry breaking hamiltonian. We have based all our discussion on the strong symmetry breaking of the form

$$\mathscr{H}'(x)=\sum_l \varepsilon_l u_l(x)\,. \tag{12.2}$$

The result for the leading divergences is that they are all accounted for (in the quark model or more generally with the commutators of CFA) by a general counterterm

$$F(\sqrt{2}E_0+E_8)=F(b_0u_0+b_8u_8+b_3u_3+b_6u_6+b_7v_7)\,. \tag{12.3}$$

In Eq. (12.3) F is unknown but b_0, b_8, b_3, b_6, and b_7 are all obtained directly from the coefficients of the algebra of chiral charges and chiral divergences in the rotated Cabibbo frame. Differently said, b_0-b_7 are the same ones calculated at lowest order. The point arises whether such a priori unexpected simplification in the problem of weak infinities also holds for other forms of $\mathscr{H}'$, for instance by also adding terms transforming as $(1,\bar{8})\oplus(\bar{8},1)$. The answer is negative. In a general theoretical frame, where one likes to consider together weak electromagnetic and strong interactions, such an answer appears as an argument in favor of the choice $(3,\bar{3})\oplus(\bar{3},3)$ for the terms breaking the chiral symmetry.

13. Higher Orders in the free Quark Model (free Quarks Coupled to W)

It is instructive to do first the 4th order (in g). We have the graphs

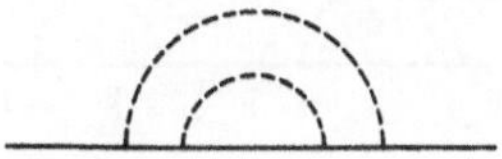
a)

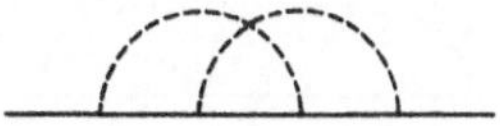
b)

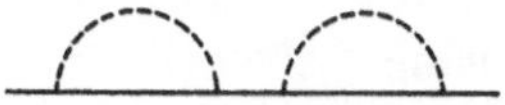
c)

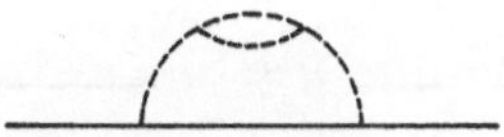
d)

b) vanishes from charge conservation. One can see this by looking at the following graph with arrows (arrows correspond to the sign of the charge):

a) and d) have "correct θ-structure": whenever there is an n propagating there is also a λ propagating. In each graph the occurring situations are of the following kind:

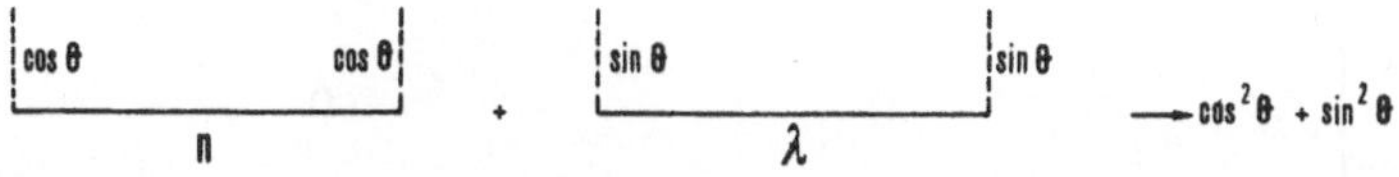

Thus everything is like in lowest order. The contributions from a and d are thus

$$\propto m_p \qquad \text{for } p\,,$$
$$\propto m_n \cos^2\theta \qquad \text{for } n\,,$$
$$\propto m_\lambda \sin^2\theta \qquad \text{for } \lambda\,.$$

Example:

intermediate state

$$n, n \to \cos^2\theta \cos^2\theta = \cos^4\theta$$
$$n, \lambda \to \qquad \cos^2\theta \sin^2\theta$$
$$\lambda, n \to \qquad \sin^2\theta \cos^2\theta$$
$$\lambda, \lambda \to \qquad \sin^4\theta$$
$$= (\cos^2\theta + \sin^2\theta)^2 = 1\,.$$

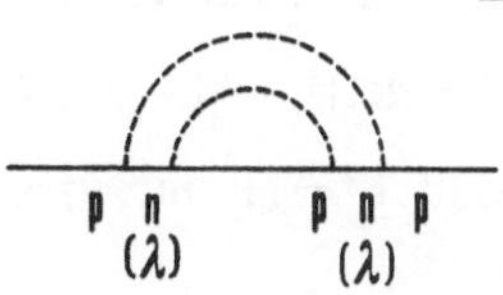

The difficult part arises with c). It looks like an improper diagrams, but it is not, because n and λ are different. So for instance one has to omit

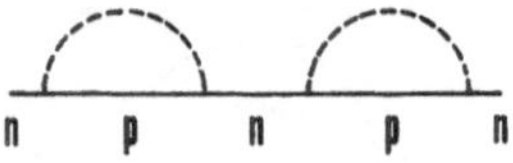

but not

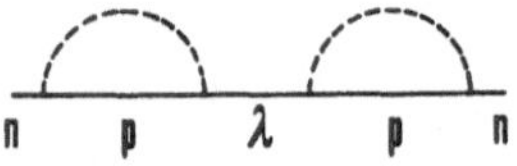

We proceed as follows. Define

C is taken to be proportional to $[Q, D^\dagger(0)] + [Q^\dagger, D(0)]$.

Precisely we write

$$[Q, D^\dagger(0)] + [Q^\dagger, D(0)] = \Sigma b_i u_i = \bar{\psi} E \psi \tag{13.1}$$

and we define

$$C = -(\sqrt{1-f_2} - 1) E \tag{13.2}$$

f_2 is the same quantity appearing in $\Sigma(k)$:

$$\Sigma(k) = \begin{bmatrix} f_2 \not{k} \frac{1+i\gamma_5}{2} & 0 & 0 \\ 0 & f_2 \not{k} \frac{1+i\gamma_5}{2} \cos^2\theta & f_2 \not{k} \frac{1+i\gamma_5}{2} \cos\theta \sin\theta \\ 0 & f_2 \not{k} \frac{1+i\gamma_5}{2} \cos\theta \sin\theta & f_2 \not{k} \frac{1+i\gamma_5}{2} \sin^2\theta \end{bmatrix}. \tag{13.3}$$

The quark propagator will be (M is the mass-matrix)

$$S_F = \frac{1}{\not{k} - M} + \frac{1}{\not{k} - M} (\Sigma - C) \frac{1}{\not{k} - M} + \cdots \tag{13.4}$$

Summing the geometric series one has

$$S^{-1} = \not{k} - M - \Sigma + C. \tag{13.5}$$

The pole location (renormalized masses) can be found from

$$\det S^{-1} = \det(\not{k} - M - \Sigma + C) = 0. \tag{13.6}$$

One finds the following result: $k^2 = m_p^2, m_n^2, n_\lambda^2$ are the poles, provided C is the one chosen above.

Now to all orders consider the strictly proper graphs at order k (this means for instance that graph d is not to be included). Define Σ_k as the sum of such graphs. Then sum up to order $2n$

$$\Sigma^{(n)} = \Sigma_2 + \Sigma_4 + \cdots + \Sigma_{2n}. \tag{13.7}$$

These graphs having the proper θ structure, one has

$$\Sigma^{(n)} = F_n \not{k} \frac{1+i\gamma_5}{2} \begin{bmatrix} 1 & 0 & 0 \\ 0 & \cos^2\theta & \cos\theta\sin\theta \\ 0 & \cos\theta\sin\theta & \sin^2\theta \end{bmatrix} \tag{13.8}$$

with

$$F_n = f_2 + f_4 + \cdots. \tag{13.9}$$

We can now choose the counterterm as

$$C_n = (1 - \sqrt{1-F_n})E = \left(\frac{1}{2} f_2 + \frac{1}{2} f_4 + \frac{1}{8} f_2^2 + \cdots\right)E. \tag{13.10}$$

Since the location of the poles at fourth order was independent of f_2, it is obvious that (the only mathematical change having been $f_2 \to F_n$) the same pole locations will be correct at any order.

We now come to the vertex corrections. We define

$$\psi = \begin{bmatrix} n \\ \lambda \end{bmatrix}, \quad \psi' = \begin{bmatrix} n' \\ \lambda' \end{bmatrix}, \quad \psi' = S\psi. \tag{13.11}$$

S being such that

$$\bar{\psi}[\not{k} - M - \Sigma + (1 - \sqrt{1-F})E]\,\psi = \bar{\psi}'(\not{k} - M)\,\psi' \tag{13.12}$$

one finds

$$S = 1 - (1 - \sqrt{1-F}) \frac{1+i\gamma_5}{2} \begin{bmatrix} \cos^2\theta & \sin\theta\cos\theta \\ \sin\theta\cos\theta & \sin^2\theta \end{bmatrix}. \tag{13.13}$$

It is easy to verify that S leaves the Cabibbo direction invariant

$$\frac{1+i\gamma_5}{2}[\cos\theta\, n' + \sin\theta\, \lambda'] = \frac{1+i\gamma_5}{2}\sqrt{1-F}\,[\cos\theta\, n + \sin\theta\, \lambda]. \tag{13.14}$$

One may wonder whether the results obtained in this section may be obtained more directly by employing the Stückelberg formalism [11] for the W boson. *T. D. Lee* [12] has in fact succeeded in summing leading and non-leading divergences in the case of a neutral W. Unfortunately the non-abelian character of the problem in the physical case of charged W brings essential complications. We wish to draw attention on a recent paper by *T. Appelquist* and *C. Carlson* [13], who have recently been able to isolate the most divergent terms in a model of weak interactions based on three vector bosons interacting with triplet currents with charges satisfying the $SU(2)$ algebra. The method used is that of the Stückelberg transformation.

IV. Leading Weak Divergences. Schemes of Current Algebra

14. Fourth Order Calculation of Leading Divergences by Current Algebra Techniques

Because of the complexity of the calculation we shall resort to the graphical technique illustrated in Section 4. We shall be interested in the extraction of the leading divergent terms from the asymptotic fourth order operator

$$M_4 = \frac{g^4}{M^4}\int \frac{d^4q}{q^2}\,\frac{d^4k}{k^2}\, q_\mu q_\nu k_\alpha k_\beta \int e^{iqx+ik(y-z)}\, T(j^+_\mu(x)\, j^+_\alpha(y)\, j^-_\beta(z)\, j^-_\nu(0)) \tag{14.1}$$

related to the graph

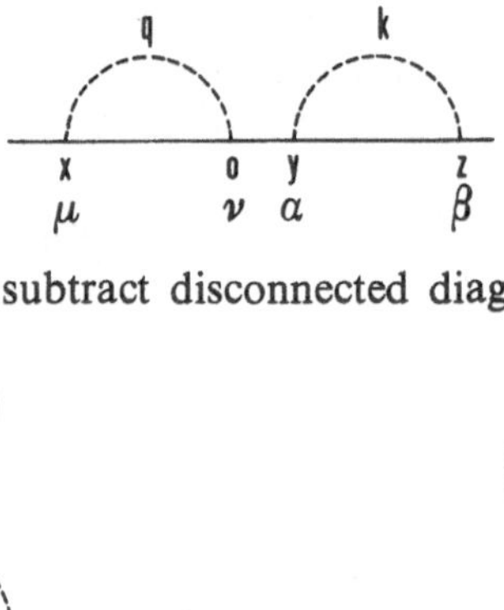

As usual one has to subtract disconnected diagrams given by graphs

The first of these graphs corresponds to a product

$$\langle 0|M_2|0\rangle \langle f|M_2|i\rangle \tag{14.2}$$

where M_2 is the second order operator. The other corresponds to

$$\langle f|i\rangle \langle 0|M_4|0\rangle . \tag{14.3}$$

The subtraction of disconnected graphs can be shown to amount to using truncated commutators, systematically, in the calculation in place of commutators (truncated commutators are commutators with their vacuum expectation values subtracted). This is true at any order. In the fourth order calculation of the present section all appearing Schwinger terms are c-numbers. Therefore the use of truncated commutators also eliminates Schwinger terms in this case. In general, at any order, Schwinger terms (of operator kind) are dealt with by use of covariant techniques. These techniques are a little more complicated and here our exposition will be more conventional to reduce any formal complication to a minimum.

We note that we have to insert one after the other (but not necessarily in the given order) the derivatives corresponding to the factors q_μ, q_ν, k_α, k_β, into the T-product in Eq. (14.1), isolating each time the contact terms that arise from the partial integration. We denote with [0] the initial graph. Thus

[O] =

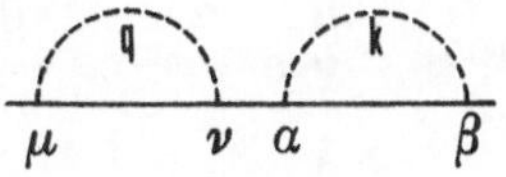

inserting q_μ (an operation that we denote as $(q_\mu \rightarrow)$) gives
$(q_\mu \rightarrow)$ [0]: [1]; [2]; [3] = 0.

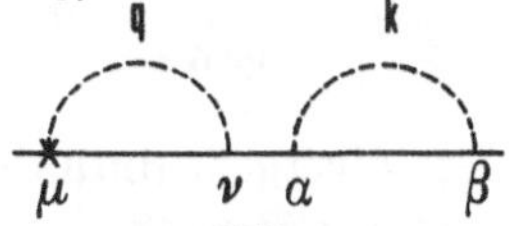

[1]

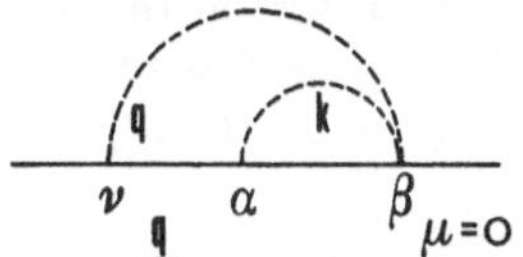

[2]

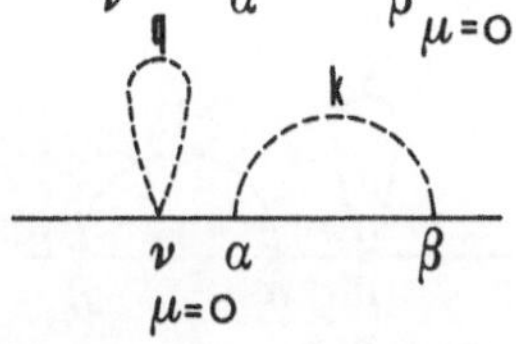

[3]

[3] vanishes as it leads to an integral odd in q. Next we operate with k_α on [1] and obtain
$(k_\alpha \rightarrow)$ [1]: [1, 1]; [1, 2] = 0 (because odd in k); [1, 3].

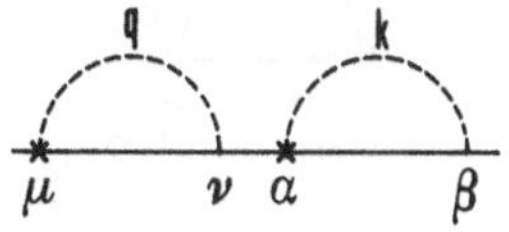

[1,1]

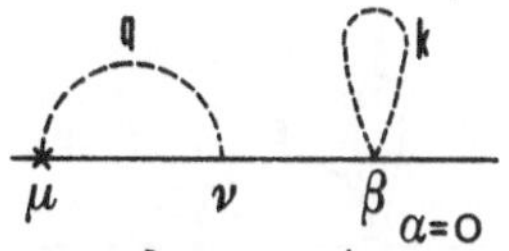

[1,2]

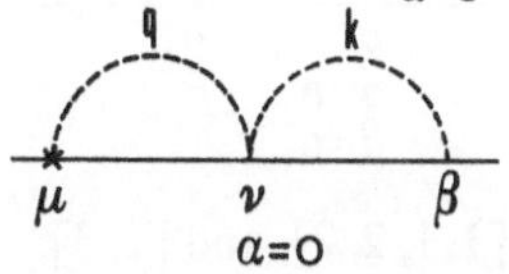

[1,3]

We are thus left with [2], [1, 1] and [1, 3]. Next, $(q_\nu \rightarrow)$ on [1, 1] gives $(q_\nu \rightarrow)$ [1, 1]: [1, 1, 1] (less divergent); [1, 1, 2]; [1, 1, 3] (less divergent).

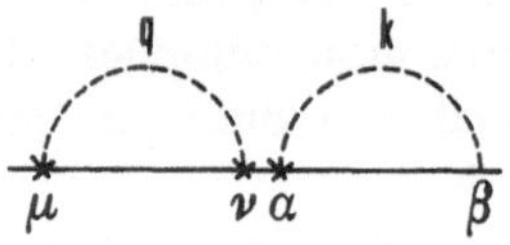

[1,1,1]

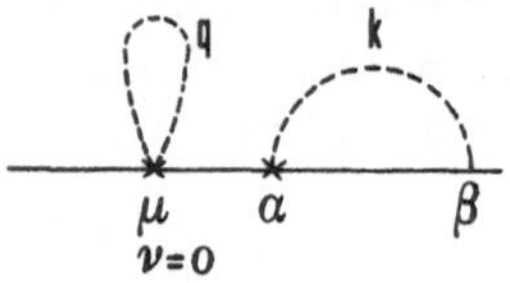

[1,1,2]

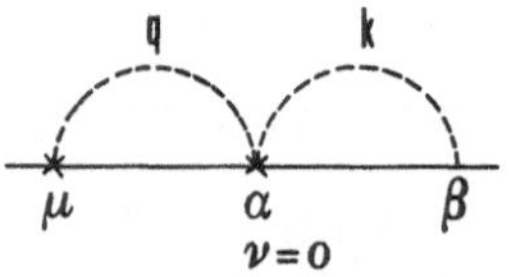

[1,1,3]

[1, 1, 1] and [1, 1, 3] are less divergent (intuitively this arises from the presence of an oscillating factor e^{iqx} in the integrand). We are left with [2], [1, 1, 2], and [1, 3]. We proceed with $(k_\beta \rightarrow)$ on [1, 1, 2] obtaining $(k_\beta \rightarrow)$ [1, 1, 2]: [1, 1, 2, 1] (less divergent); [1, 1, 2, 2]; [1, 1, 2, 3] (less divergent).

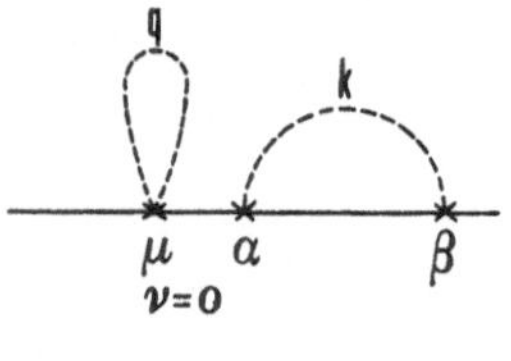

[1,1,2,1]

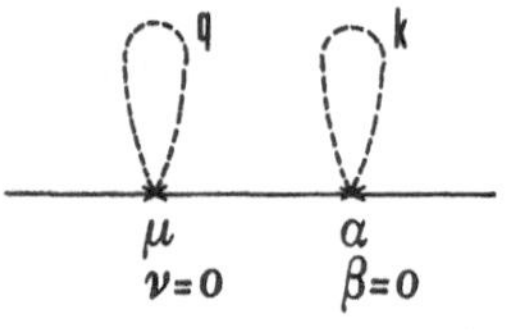

[1,1,2,2]

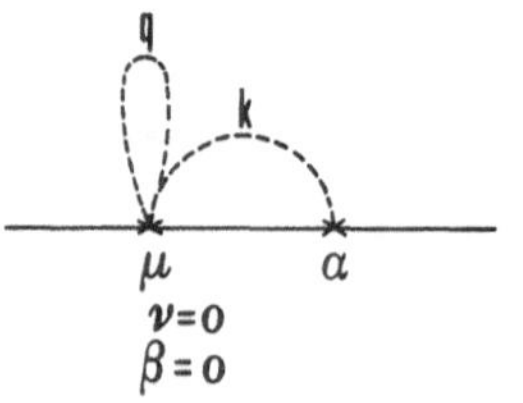

[1,1,2,3]

We are now left with [2], [1, 1, 2, 2], and [1, 3].

On [1, 3] we still have to operate with q_ν and k_β. After $(k_\beta \rightarrow)$ one has $(k_\beta \rightarrow)$ [1, 3]: [1, 3, 1]; [1, 3, 2]; [1, 3, 3].

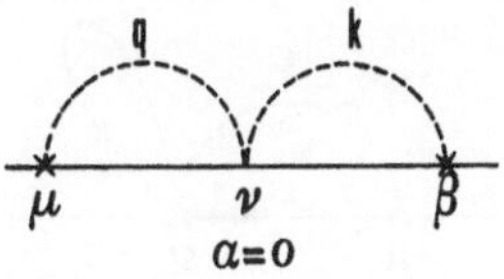

[1,3,1]

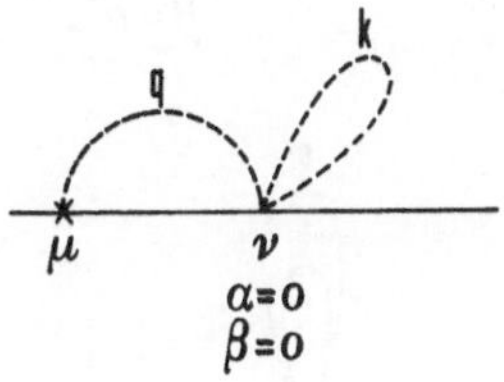

[1,3,2]

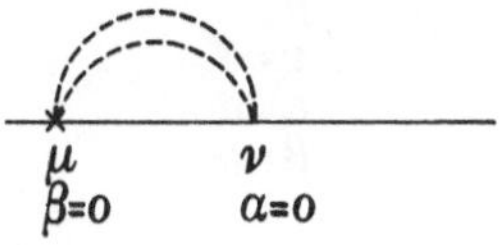

[1,3,3]

Next $(q_\nu \rightarrow)$ on [1, 3, 1] gives
$(q_\nu \rightarrow)$ [1, 3, 1]: [1, 3, 1, 1] (less divergent); [1, 3, 1, 2] (less divergent); [1, 3, 1, 3] (less divergent).

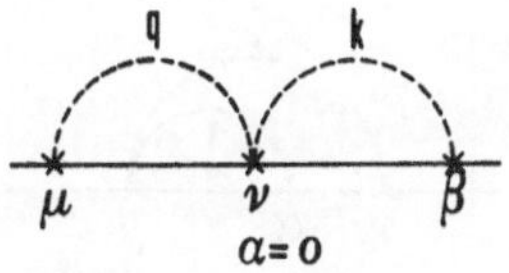

[1,3,1,1]

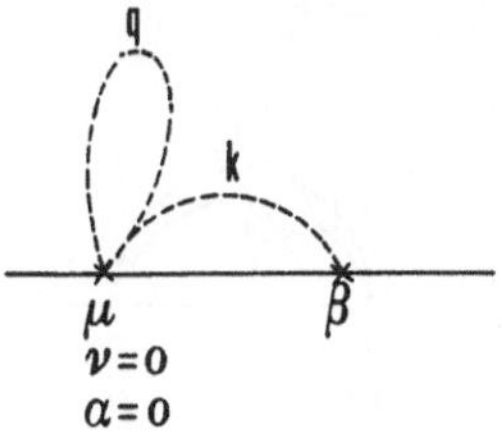

[1,3,1,2]

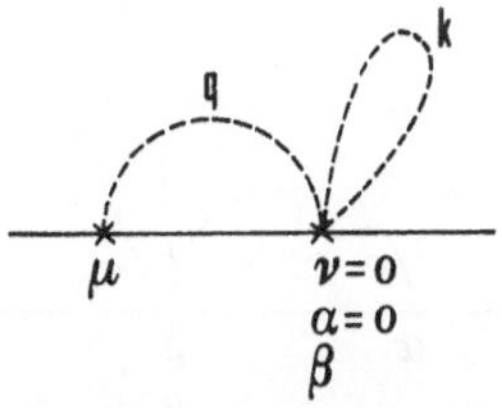

[1,3,1,3]

[1, 3, 1] gives therefore only less divergent terms.

$(q_\nu \rightarrow)$ on [1, 3, 2] gives
$(q_\nu \rightarrow)$ [1, 3, 2]: [1, 3, 2, 1] (less divergent); [1, 3, 2, 2].

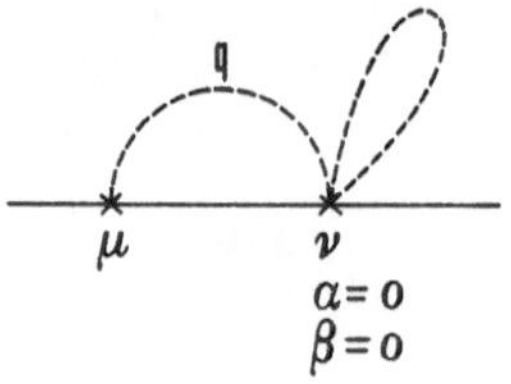

[1,3,2,1]

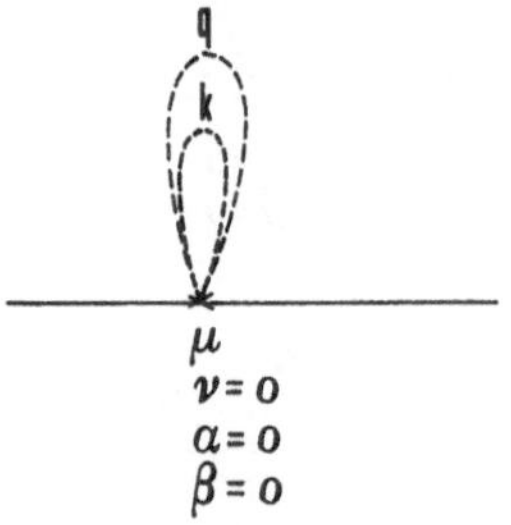

[1,3,2,2]

We are thus left at this stage, with [2], [1, 1, 2, 2], [1, 3, 2, 2], and [1, 3, 3]. Let us now reduce [2]. We have to insert q_ν, k_α, and k_β. Applying q_ν gives
$(q_\nu \rightarrow)$ [2]: [2, 1]; [2, 2]; [2, 3].

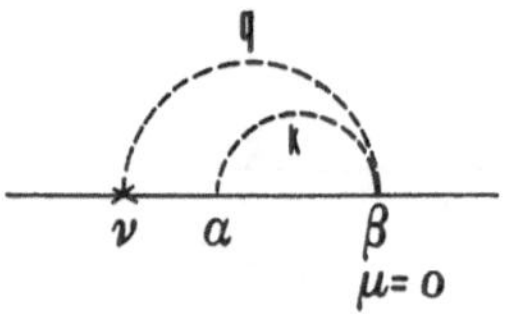

[2,1]

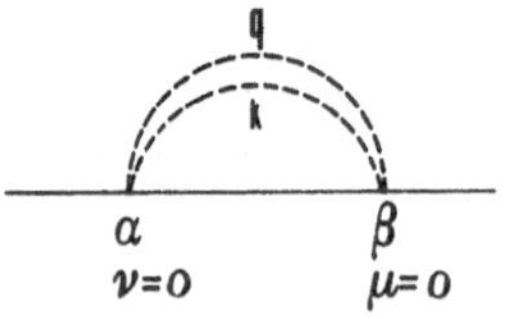

[2,2]

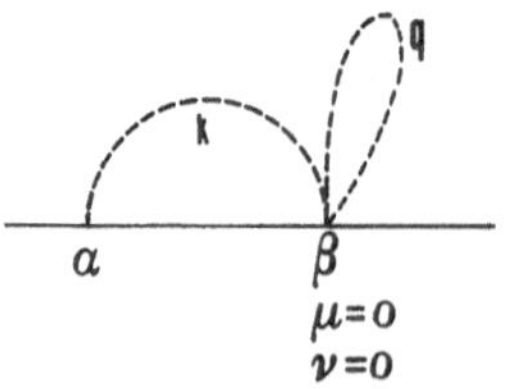

[2,3]

[2, 1] is similar to [1, 3]. It gives similar terms as [1, 3]. On [2, 3] we operate with k_α and get
($k_\alpha \rightarrow$) [2, 3]: [2, 3, 1]; [2, 3, 2] = 0 (because odd in k).

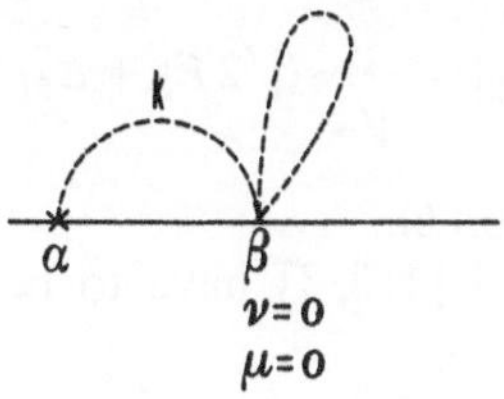

[2,3,1]

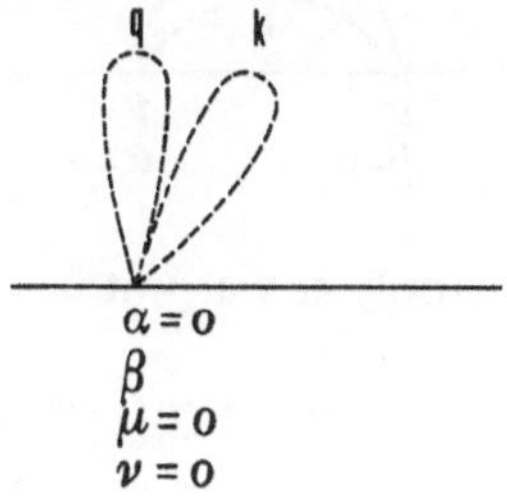

[2,3,2]

[2, 3, 2] vanishes after integrating on k (we have still a factor k_β not "used"); [2, 3, 1] can be compared to [1, 3, 2] and seem to give similar terms.

At this stage we are left with [2, 2], [1, 1, 2, 2], [1, 3, 2, 2], and [1, 3, 3]. Graphs of the type [1, 1, 2, 2] will clearly give an integral of the type

$$\int \frac{d^4q d^4k}{q^2k^2} \int d^4x \, T\{[Q^-, D^+(x)] \, [Q^-, D^+(0)] + [Q^+, D^-(x)] \, [Q^+, D^-(0)]\} \tag{14.4}$$

(Q and D are the rotated chiral charges and currents, called Q' and D' in Section 6). This is an iteration of the second order leading divergence. Graphs of the type [1, 3, 2, 2] will give contributions of the type

$$\{[Q^+, D^-] + [Q^- D^+]\} \int \frac{d^4q d^4k}{q^2k^2} \tag{14.5}$$

and

$$\{[Q^3 [Q^-, D^+(0)]] - [Q^3 [Q^+, D^-(0)]]\} \int \frac{d^4q d^4k}{q^2k^2} \tag{14.6}$$

according to the order in performing the operations. From the algebra of chiral currents and divergences one sees that such terms are all proportional to $\sqrt{2}E_0+E_8$. In fact

$$[Q^{\mp}, D^{\pm}] = \mp D_3 + \frac{i}{\sqrt{3}}(\sqrt{2}E_0+E_8), \tag{14.7}$$

$$[Q_3, D_3] = \frac{i}{\sqrt{3}}(\sqrt{2}E_0+E_8) \tag{14.8}$$

(E_i were called E_i' in Section 6).

The graphs [2, 2] and [1, 3, 3] have to be reduced further. Let us first consider [1, 3, 3]

[1,3,3]

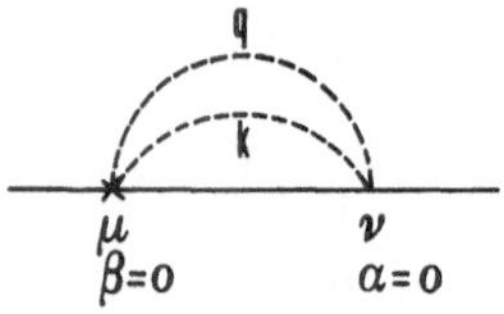

It is an integral of the type (only q_ν was not used)

$$\int \frac{d^4q\,d^4k}{q^2k^2} q_\nu \int d^4z e^{-i(q-k)z} T\{[Q^+, D^-(z)]\, j_\nu^3(0)\}\,. \tag{14.9}$$

The essential point is that one cannot directly insert q_ν by usual partial integration technique, because the momentum in the exponential is not q but $q-k$. Let us call $q-k=p$. We decompose

$$q_\mu = q_\mu - \frac{pq}{p^2}p_\mu + \frac{pq}{p^2}p_\mu = \frac{pq}{p^2}p_\mu = \frac{pq}{p^2}p_\mu + \frac{1}{p^2}(p_\lambda q_\mu - q_\lambda p_\mu)p_\lambda. \tag{14.10}$$

The (longitudinal) term $(pq)\,p_\mu/p^2$ in front of the exponential e^{ipx} allows the usual partial integration technique. The reduction technique applied to this term gives terms of the kind we have already met. Let us deal with

$$\frac{1}{p^2}(p_\lambda q_\mu - q_\lambda p_\mu)\, p_\lambda \int d^4z e^{-ipz} T([Q^+, D^-(z)]\, j_\mu^3(0))\,. \tag{14.11}$$

As usual, we have for this integral the result

$$\begin{aligned} &-\frac{i}{p^2}(p_\lambda q_\mu - q_\lambda p_\mu)\int d^4z e^{-ipz} \\ &\cdot\left\{\delta_{\lambda 0}\delta(z_0)\,[[Q^+, D^-(0)], j_\mu^3(z)] + T\left([Q^+, D^-(0)]\frac{\partial j_\mu^3(z)}{\partial z_\lambda}\right)\right\}. \end{aligned} \tag{14.12}$$

The last term is clearly less divergent. We rewrite the first term as

$$-\frac{i}{p^2}p_\lambda q_\mu \int d^4z e^{-ipz}\delta(z_0)\,[[Q^+, D^-(0)], (\delta_{\lambda 0} j_\mu^3(z) - \delta_{\mu 0} j_\lambda^3(z))]\,. \tag{14.13}$$

Now $\delta_{\lambda 0} j^3_\mu - \delta_{\mu 0} j^3_\lambda$ has only components proportional to j_i. In fact, for $\mu=0$, $\lambda=i$ one has $-j^3_i$; for $\mu=i$, $\lambda=0$ one has j^3_i, for $\mu=0$, $\lambda=0$ one has $j^3_0 - j^3_0 = 0$; for $\mu=i$, $\lambda=i$ one has zero. But the commutators $[[Q^+, D^-(0)]\, j_i(z)]$ can be seen to vanish from the field algebra. We must next discuss [2, 2]. This graph is responsible for the appearance of additional commutators in the final result.

[2,2]

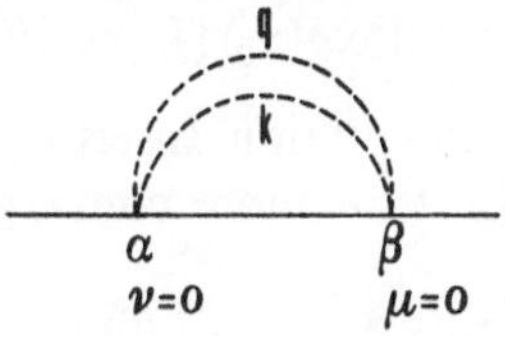

It gives rise to an integral

$$\int \frac{d^4 q\, d^4 k}{q^2 k^2} k_\alpha k_\beta \int d^4 z\, e^{i(q-k)z}\, T(j^3_\alpha(z)\, j^3_\beta(0)) . \tag{14.14}$$

Calling

$$\begin{aligned} q-k &= Q , \\ q+k &= p , \end{aligned} \tag{14.15}$$

we have in the integral a product of the type

$$\begin{aligned} &(Q-p)_\alpha (Q-p)_\beta e^{iQz} \\ &= Q_\alpha Q_\beta e^{iQz} - (p_\alpha Q_\beta + Q_\alpha p_\beta) e^{iQz} + p_\alpha p_\beta e^{iQz} \\ &= -\partial_\alpha \partial_\beta e^{iQz} + i(p_\alpha \partial_\beta + \partial_\alpha p_\beta) e^{iQz} + p_\alpha p_\beta e^{iQz} . \end{aligned} \tag{14.16}$$

In the last sum only the terms $p_\alpha p_\beta e^{iQz}$ need a new kind of treatment. Terms of the kind $\partial_\alpha \partial_\alpha e^{iQz}$ have already been encountered. Terms $p_\alpha \partial_\beta + \partial_\alpha p_\beta$ can be similarly treated. For all these terms one goes on by partial integrating and finally arrives to leading divergences of the type we have already found. A more complex procedure has to be developed for the term

$$p_\mu p_\nu \int e^{iQx}\, T(j_\mu(x)\, j_\nu(o))\, d^4 x \tag{14.17}$$

(we have omitted the index 3 to simplify the notations; also for uniformity we shall use the symbol q in place of Q in Eq. (14.17)) to usual we decompose p_μ into its longitudinal and transverse part with respect to q_μ and write the identity

$$\begin{aligned} p_\mu p_\nu = \frac{1}{q^4} [&(q_\lambda p_\mu - p_\lambda q_\mu)(q_\varrho p_\nu - q_\nu p_\varrho)\, q_\lambda q_\varrho \\ &+ (q_\lambda p_\mu - p_\lambda q_\mu)(pq)\, q_\lambda q_\nu \\ &+ (q_\varrho p_\nu - q_\nu p_\varrho)(qp)\, q_\mu q_\varrho + (qp)^2 q_\mu q_\nu] . \end{aligned} \tag{14.18}$$

The last three terms in Eq. (14.18) allow for further partial integrations, as they contain factors q_ν, q_μ, and $q_\mu q_\nu$ respectively. For instance

$$q_\mu \int e^{iqx} T(j_\mu(x)\, j_\nu(0))\, d^4x$$

can be reduced to

$$\int e^{iqx} d^4x\, T(\mathcal{D}(x), j_\nu(0)) \tag{14.19}$$

plus a term

$$\int e^{iqx} d^4x\, \delta(x_0)\, [j_0(x), j_\nu(0)]$$

which vanishes by partial integration. Inserting the factor q_ϱ the terms in Eq. (14.19) gives now rise to a more convergent term containing the product

$$T\left(\mathcal{D}(x), \frac{\partial j_\nu(0)}{\partial x_\varrho}\right)$$

plus a term containing

$$\delta_{\varrho 0}[\mathcal{D}(x), j_\nu(0)]\, \delta(x_0)\,. \tag{14.20}$$

The latter term, Eq. (14.20), is of the usual kind $[\mathcal{D}(x), j_0(0)]\, \delta(x_0)$ since for $\nu = i$ one has $[\mathcal{D}(x), j_i(0)]\, \delta(x_0) = 0$. We have now to consider the first term in Eq. (14.18).

We shall reduce further the product

$$(q_\lambda p_\mu - p_\lambda q_\mu)(q_\varrho p_\nu - q_\nu p_\varrho)\, q_\lambda q_\varrho \int e^{iqx} T(j_\mu(x)\, j_\nu(0))\, d^4x\,. \tag{14.21}$$

Inserting q_λ we have

$$\begin{aligned}(14.21) = {} & i(q_\lambda p_\mu - p_\lambda q_\mu)(q_\varrho p_\nu - q_\nu p_\varrho)\, q_\varrho \\ & \times \int d^4x\, e^{iqx} \left\{ \delta_{\lambda 0} [j_\mu(x), j_\nu(0)]\, \delta(x_0) + T\left(\frac{\partial j_\mu(x)}{\partial x_\lambda}\, j_\nu(0)\right)\right\}.\end{aligned} \tag{14.22}$$

The first term in Eq. (14.22) vanishes after integration on q. One has, recalling our remarks on Schwinger terms.

$$\int d^4x\, e^{iqx} \delta_{\lambda 0} [j_\mu(x), j_\nu(0)]\, \delta(x_0) = \delta_{\lambda 0} O_{\mu\nu}(0)$$

where $O_{\mu\nu}$ is some operator; after integrating on q the integral vanishes because of odd integrand. We are left with

$$\begin{aligned} & i(q_\lambda p_\mu - p_\lambda q_\mu)(q_\varrho p_\nu - p_\varrho q_\nu)\, q_\varrho \int e^{iqx} d^4x\, T\left(\frac{\partial j_\mu(x)}{\partial x_\lambda}\, j_\nu(0)\right) \\ & = -(q_\lambda p_\mu - p_\lambda q_\mu)(q_\varrho p_\nu - p_\varrho q_\nu) \int e^{-iqx} d^4x \left\{ \delta_{\varrho 0} \left[\frac{\partial j_\mu(0)}{\partial x_\lambda}, j_\nu(x)\right] \delta(x_0) \right. \\ & \quad \left. + T\left(\frac{\partial j_\mu(0)}{\partial x_\lambda}, \frac{\partial j_\nu(x)}{\partial x_\varrho}\right)\right\}. \end{aligned} \tag{14.23}$$

In Eq. (14.23) the term containing the T-product is more convergent. By repeated use of the antisymmetry of the product of momenta in front of the integral we can rewrite the remaining expression as as

$$q_\lambda p_\mu q_\nu p_\varrho \int d^4x e^{iqx} \delta(x_0) \left[\frac{\partial j_\mu(x)}{\partial x_\lambda} - \frac{\partial j_\lambda(x)}{\partial x_\mu}, \delta_{\varrho 0} j_\nu(0) - \delta_{\nu 0} j_\varrho(0) \right]. \tag{14.24}$$

The combination

$$\delta_{\varrho 0} j_\nu - \delta_{\nu 0} j_\varrho$$

is non vanishing only if $\nu = j$, $\varrho = 0$ (or viceversa) and in this case is proportional to j_i. The relevant commutators are thus

$$\delta(x_0) \left[\frac{\partial j_\mu(x)}{\partial x_\lambda} - \frac{\partial j_\lambda(x)}{\partial x_\mu}, j_i(0) \right].$$

With field algebra commutators this expression vanishes for μ, λ both spacelike. The only appearing commutator is thus

$$\delta(x_0) \left[\frac{\partial j_0}{\partial x_i} - \frac{\partial j_i}{\partial x_0}, j_l \right]. \tag{14.25}$$

This completes our proof that the leading divergences at fourth order only depend on charge-divergence commutators and on commutators of the kind in Eq. (14.25).

15. Compound Field Algebra (CFA) and Leading Weak Divergences

We recall briefly the main developments leading to field algebra. We first recall, for a most direct introduction, the concept of field-current identy by *Kroll*, *Lee* and *Zumino* [14]. The concept originates from a rigorous Lagrangian scheme of vector dominance which includes consistently gauge invariance. For simplicity let us limit ourselves to the ϱ-meson. The Lagrangian density is assumed to be

$$-\frac{1}{2} m^2 \varrho_\mu \varrho_\mu + L' \tag{15.1}$$

where L' is invariant under a substitution

$$\varrho_\mu \to \varrho_\mu + \frac{1}{g} \partial_\mu \Lambda \tag{15.2a}$$

and a corresponding transformation on the "matter fields" ψ, which occur in L'. One introduces electromagnetism through the substitution

$$\varrho_\mu \to \varrho_\mu + \frac{e}{g} A_\mu \tag{15.2b}$$

inside L'. Of course the treatment in this example is only limited to isovector photons. This prescription clearly respects gauge invariance as one immediately verifies. One writes down the equations of motion

$$\frac{\partial F_{\lambda\mu}}{\partial x_\lambda} + \frac{e}{g} \frac{\delta L'}{\delta \varrho_\mu} = 0 \tag{15.3}$$

and

$$-m^2 \varrho_\mu + \frac{\delta L'}{\delta \varrho_\mu} = 0 \tag{15.4}$$

directly from the Lagrangian (15.1) with A_μ introduced through Eq. (15.2). Eqs. (15.3) and (15.4) together exhibit field-current identity in the form

$$j_\mu^{(\mathrm{em})} = \frac{\partial F_{\lambda\mu}}{\partial x_\lambda} = -\frac{em^2}{g} \varrho_\mu . \tag{15.5}$$

One can now perform a wave-function renormalization by introducing the renormalized field (to be called again ϱ_μ, whereas ϱ_μ^0 is the unrenormalized field)

$$\varrho_\mu^0 = Z^{\frac{1}{2}} \varrho_\mu \tag{15.6}$$

and one also introduces

$$Z_0 = \left(\frac{m}{m_0}\right)^2 \tag{15.7}$$

the ratio between renormalized and unrenormalized masses. One can easily calculate the renormalized and unrenormalized sources J_ν and J_ν^0 respectively. They are related by

$$J_\nu^0 = J_\nu + (1 - Z_0) \frac{1}{g} \frac{\partial G_{\mu\nu}}{\partial x_\mu} = -\frac{m^2}{g}\left(\varrho_\nu - \frac{1}{m_0^2} \frac{\partial G_{\mu\nu}}{\partial x_\mu}\right) \tag{15.8}$$

where g is the renormalized coupling and $G_{\mu\nu}$ is the field tensor for the ϱ-meson. In writing Eq. (15.8) we have conventionally defined the unrenormalized coupling as

$$g_0 = g Z^{\frac{1}{2}} Z_0^{-1} . \tag{15.9}$$

From Eq. (15.8) one makes the important observation that for $m_0 \to \infty$ the e.m. current (which is here identical to the field) also becomes identical to the *unrenormalized* current J_ν^0. This reminds us of the current-

current identity proposed by *Gell-Mann* and *Zachariasen* [15]. In the above model the spectral sum rules are a direct consequence of the canonical commutation relations and of the field equations. The extension to a non-abelian situation [16] (SU_2, $SU_2 \times SU_2$, SU_3, $SU_3 \times SU_3$) presents a formal difficulty connected to the occurrence in commutators of fields and their time derivatives of ambiguous terms proportional to bilinear expressions in the gauge fields taken at the same space-time point. Such products are not well-defined, although a formal argument based on relativistic invariance, suggests that they do not contribute between vacuum states (and therefore they do not appear in the usual spectral sum rules). Furthermore in the non-abelian case (except SU_2) some currents are non-conserved. The non abelian case is the one of interest here (we are dealing with chiral $SU_3 \times SU_3$). It leads to the commutators of *Lee*, *Weinberg*, and *Zumino* which can be directly constructed out of the field equations and the canonical commutators. For the fields $\Phi_{a\mu}$ (where the index a refers to the internal symmetry) one assumes a Yang-Mills Lagrangian density

$$L = -\tfrac{1}{4} F_{a\mu\nu} F_{a\mu\nu} - \tfrac{1}{2} m_0^2 \Phi_{a\mu} \Phi_{a\mu} + L'(\psi, D_\mu \psi)\,, \tag{15.10}$$

$$F_{a\mu\nu} = \partial_\mu \Phi_{a\nu} - \partial_\nu \Phi_{a\mu} + g_0 C_{abc} \Phi_{b\mu} \Phi_{c\nu}\,, \tag{15.11}$$

where C_{abc} are the structure constants of the internal symmetry algebra and T_a is the representation of the algebra on ψ. The field equations are

$$m_0^2 \Phi_{a\nu} = \frac{\partial}{\partial x_\mu} F_{a\mu\nu} + g_0 J_{a\nu}\,, \tag{15.12}$$

$$J_{a\nu} = F_{b\nu\mu} C_{abc} \Phi_{c\mu} + \frac{\partial L'}{\partial (D_\nu \psi)} T_a \psi\,. \tag{15.13}$$

It follows that

$$\frac{\partial \Phi'_{a\nu}}{\partial x_\nu} = \frac{m_0^2}{g_0} \frac{\partial \ {}_{a\nu}}{\partial x_\nu} = \frac{\partial J_{a\nu}}{\partial x_\nu} \tag{15.14}$$

if

$$\Phi'_{a\nu} = \frac{m_0^2}{g_0} \Phi_{a\nu} \tag{15.15}$$

so that if $J_{a\nu}$ is conserved so is $\Phi'_{a\nu}$ and viceversa. The equal time commutators are easily obtained. Clearly

$$\delta(x_0)\,[\Phi_{ai}(x), \Phi_{bj}(0)] = 0\,. \tag{15.16}$$

Also

$$\pi_{ai}(x) = \frac{\partial L(x)}{\partial \dot{\Phi}_{ai}} = -i F_{ai4}(x)\,.$$

Therefore

$$\begin{aligned}
&\delta(x_0)\,[\Phi_{ai}(x), F_{bj4}(0)] = -\delta_{ab}\delta_{ij}\delta(x)\\
&-\delta_{ab}\delta_{ij}\delta(x) = \delta(x_0)\,[\Phi_{ai}(x), \partial_j\Phi_{b4}(0) - \partial_4\Phi_{bj}(0)]\\
&+g_0\delta(x_0)\,C_{bcd}\Phi_{lj}(0)\,[\Phi_{ai}(x), \Phi_{d4}(0)]\,.
\end{aligned} \tag{15.17}$$

The last commutator in Eq. (15.17) can be calculated using the equations of motions (for $v=4$) in the form

$$m_0\Phi_{a4} = i\frac{\partial}{\partial x_j}\pi_{aj}(x) - ig_0\pi_{bj}(x)\,C_{abc}\Phi_{cj}(x) + i\pi_\psi(x)\,T_a\psi(x)\,. \tag{15.18}$$

Thus

$$\delta(x_0)m_0^2[\Phi_{a4}(x), \Phi_{bi}(0)] = \delta_{ab}\frac{\partial}{\partial x_i}\delta(x) - g_0C_{abc}\Phi_{ci}(0)\,\delta(x) \tag{15.19}$$

where we have assumed that

$$[\pi_\psi(x)\,T_a\psi(x), \Phi_{bi}(0)] = 0\,. \tag{15.20}$$

We therefore find

$$\begin{aligned}
&\delta(x_0)\,[\Phi_{ai}(x), \partial_j\Phi_{b4}(0) - \partial_4\Phi_{bj}(0)]\\
&= -\delta_{ab}\delta_{ij}\delta(x) - \frac{g_0}{m_0^2}\,C_{bla}\Phi_{lj}(0)\frac{\partial}{\partial x_i}\delta(x)\\
&-\left(\frac{g_0}{m_0}\right)^2 C_{bld}C_{dac}\Phi_{lj}(0)\,\Phi_{ci}(0)\,\delta(x)\,.
\end{aligned} \tag{15.21}$$

Finally from Eq. (15.18) one obtains

$$\delta(x_0)\,m_0^4[\Phi_{a4}(x), \Phi_{b4}(0)] = ig_0C_{abc}m_0^2\Phi_{c4}(x)\,\delta(x)\,. \tag{15.22}$$

Collecting the relevant results and introducing $\Phi' = \frac{m_0^2}{g_0}\Phi$ (to be identified with the currents) one finally obtains

$$\delta(x_0)\,[\Phi'_{ai}(x), \Phi'_{bj}(0)] = 0\,, \tag{15.23}$$

$$\begin{aligned}
&\delta(x_0)\,[\Phi'_{a4}(x), \Phi'_{bi}(0)]\\
&= -C_{abc}\Phi'_{ci}(0)\,\delta(x) - \frac{Z_0}{Z}\left(\frac{m}{g}\right)^2\delta_{ab}\frac{\partial}{\partial x_i}\delta(x)\,,
\end{aligned} \tag{15.24}$$

$$\begin{aligned}
&\delta(x_0)\,[\partial_4\Phi'_{aj}(x) + \partial_j\Phi'_{a4}(x), \Phi'_{bj}(0)]\\
&= -\frac{1}{Z}m^2\left(\frac{m}{g}\right)^2\delta_{ab}\delta_{ij}\delta(x) + C_{abc}\Phi'_{cj}(0)\frac{\partial}{\partial x_i}\delta(x)\\
&-\left(\frac{g}{m}\right)^2\frac{Z}{Z_0}\,C_{adc}C_{bde}\Phi'_{cj}(0)\,\Phi'_{ei}(0)\,\delta(x)\,,
\end{aligned} \tag{15.25}$$

$$\delta(x_0)\,[\Phi'_{a4}(x), \Phi'_{b4}(0)] = iC_{abc}\Phi'_c(0)\,\delta(x)\,. \tag{15.26}$$

We have introduced

$$Z = m^2 \left(\frac{m}{g}\right)^2 \left(\frac{g_0}{m_0}\right)^2 \frac{1}{m_0^2}, \tag{15.27}$$

$$Z_0 = \frac{m^2}{m_0^2} \tag{15.28}$$

as given above.

In addition to the currents our set of field quantities includes the u's and v's. A consistent way of including such operators is to introduce independent canonical fields s and p, of scalar and pseudoscalar character, such that

$$s_i = u_i - \langle u_i \rangle_0, \tag{15.29}$$

$$p_i = v_i. \tag{15.30}$$

By construction of an explicit lagrangian, including the symmetry breaking in the postulated form of Eq. (7.1), one can verify that the canonical rules for s_i and p_i are in fact consistent. The extension of field algebra (to be abbreviated as F.A.) obtained in this way will be called extended field algebra (to be abbreviated as E.F.A.). This is the algebraic frame in which we work. By a straightforward, although rather lengthy, counting argument one can show, for the fourth order term discussed in this section, that the introduction of a counterterm proportional to

$$\sqrt{2} E_0 + E_8 \tag{15.31}$$

in fact takes into account *all* the leading divergences except for the terms proportional to the commutator in Eq. (14.25), which would spoil such result. It is therefore important to establish under what conditions the physically important situation of leading divergences again behaving as terms of $(3, \bar{3}) \oplus (\bar{3}, 3)$ (as for example in the quark model of section 13) obtains. To this purpose we shall concentrate on a limiting case of field algebra: compound field algebra, to be abbreviated as CFA.

One can define compound field algebra (CFA) by a formal limiting procedure on the bare quantities m_0 and g_0 [17]. Alternatively one can use Eqs. (15.27) and (15.28) to perform the limit on Z and Z/Z_0. For CFA the limit is

$$\frac{m_0^2}{g_0^2} \to \infty, \left(\frac{m_0}{g_0}\right)^2 m_0^2 \to \infty. \tag{15.32}$$

In terms of Z and Z/Z_0

$$\frac{Z}{Z_0} \to 0, \quad Z \to 0. \tag{15.33}$$

The Feynman propagator $\Delta'_{\mu\nu}(q^2)$ for the gauge particle is infinite

$$\Delta'_{\mu\nu}(0) = \infty \, . \tag{15.34}$$

In the commutation relation $[j_0(x), j_i(0)]\,\delta(x_0)$, see Eq. (15.24), the Schwinger term is infinite, the c-number term in the commutator $[\partial_0 j_i(x) - \partial_i j_0(x), j_k(o)]\,\delta(x_0)$, see Eq. (15.25), is also infinite. The vanishing of Z/Z_0 suggests that the last term in the commutator of Eq. (15.25) can be neglected.

The latter commutator is that appearing in the (logarithmically divergent) electromagnetic mass differences. Of the three terms appearing on the right-hand-side of Eq. (15.25) only the q-number term proportional to the δ-function will clearly contribute to the mass-differences (the other terms are: one a c-number, the other a Schwinger term). In general one would expect logarithmically divergent e.m. mass differences (that is uncalculable in this context). However if the last term in Eq. (15.25) is absent the e.m. mass differences are finite [17]. From the results of section 14 it is apparent that in CFA the leading divergences up to fourth order are proportional to commutators of charge and currents. For the higher orders a formal proof can be given subject to some qualifications. To this end we note that already in dealing with Eq. (15.25) one is faced by a product of two field operators taken at the same space-time point. This product is not well defined in field theory. A second important point is that the commutator algebra in the limit $Z/Z_0 \to 0$ and $Z \to 0$ does not satisfy the Jacobi identity. To verify this one has simply to apply the Jacobi identity to $[\Phi_0^{\alpha}(x)\,[\dot{\Phi}_i^{\beta}(y), \Phi_j^{\gamma}(z)]]\,\delta(y_0 - z_0)\,\delta(x_0 - y_0)$. The equal time commutator $[\dot{\Phi}_i, \Phi_j]$ has (before the limit) a term proportional to Z/Z_0. Such a term commuted again with Φ_0 gives rise to some terms independent of Z and Z_0, because of the inverse factor Z_0/Z appearing in the last term of Eq. (15.24). On the other hand the Jacobi identity holds (before the limit). But then it cannot hold in the limit because the term we have discussed will not be there if the limit is performed first (the equal time commutators $[\dot{\Phi}_i[\Phi_j, \Phi_0]]$ and $[\Phi_j[\Phi_0, \dot{\Phi}_i]]$ do not contain any finite term from cancellation of Z/Z_0 with Z_0/Z). The non-validity of the Jacobi identity makes it necessary to formulate proofs by first calculating the leading divergences outside the limit and only afterwards performing the limit to CFA. This is in fact what we have done in the 4-th order. However, the proof under such conditions becomes extremely complicated at the higher orders.

By a simple combinatorial argument that will not be reproduced here one can show, from our fourth order results, that for CFA (enlarged to extended "compound field algebra" ECFA, to include the densities u and v) one has the result that a counterterm proportional to $\sqrt{2}E'_0 + E'_8$

renormalizes the leading divergences of the theory. By a complex proof in EFA and then in the limit of ECFA one can show that the same results hold at all orders (see following sections). We have already mentioned some of the qualifications to be made before enunciating such a result. Also in the higher order result they are all connected with the question of proper definition of field products at same space time point. Nevertheless it is astonishing how the formal proof goes through in spite of its great algebraic complexity and apparently suggests that the result has a definite validity. On the other hand the result is consistent with that obtained in the quark model. Before concluding this section we can point out that CFA gives a definite prediction for the asymptotic behaviour of the total cross-section for electron-positron annihilation into hadrons. For a discussion of this point and of the connected question of Weinberg sum rules we refer to a recent summary talk [18], where other limiting cases of field algebra including the Sugawara model are examined.

16. Leading Divergences at Higher Orders

We shall summarize here the main steps of the formal result we have stated in the previous section, namely that a counterterm proportional to

$$\sqrt{2}E'_0 + E'_8 = b_0 u_0 + b_3 u_3 + b_8 u_8 + b_6 u_6 + b_7 v_7 ,$$

where the coefficients $b_0 \dots b_7$ are known functions of the ε's and of θ, is sufficient to renormalize the leading divergences, for an $SU_3 \times SU_3$ breaking belonging to $(3, \overline{3}) \oplus (\overline{3}, 3)$, and if the commutation rules of ECFA (extended compound field algebra) hold. We have already described the formal character of the proof, which is carried out by explicitly allowing formal manipulations with products of operators, which in fact are not well-defined. We have not yet succeded in specifying the limits of validity of the result, as implied by the above difficulty. On the other hand the difficulty mentioned in the previous section, related to the necessity of carrying out the proof first in EFA (extended field algebra) and by going afterwards to ECFA by carrying out the relevant limit, can be bypassed by the formal argument developed below. In view of the algebraic complexity of the argument we shall be compelled to resort to a rather abstract language and will not give the proofs in detail. The developments below go back to unpublished work in collaboration with *G. Sartori* and *M. Tonin*. Before presenting the argument we want again to warn the reader about its incompleteness because of the lack of rigour in treating products of local fields.

We start from the Lagrangian (in unrenormalized form)

$$\begin{aligned} L = & -\tfrac{1}{4}G_{\mu\nu}^{(0)\alpha}G_{\mu\nu}^{(0)\alpha} - \tfrac{1}{2}m_0^2\phi_\mu^{(0)\alpha}\phi_\mu^{(0)\alpha} - \tfrac{1}{2}D_\mu^{(0)\alpha\beta}u^{(0)\beta}D_\mu^{(0)\alpha\gamma}u^{(0)\gamma} \\ & -\tfrac{1}{2}\mu_0^2 u^{(0)\alpha}u^{(0)\alpha} + \tfrac{1}{3}a_0^{\alpha\beta\gamma}u^{(0)\alpha}u^{(0)\beta}u^{(0)\gamma} \\ & +\tfrac{1}{4}b_0^{\alpha\beta\gamma\delta}u^{(0)\alpha}u^{(0)\beta}u^{(0)\gamma}u^{(0)\delta} \\ & +G_0\overline{\psi}^{(0)}T^\alpha\psi^{(0)}u^{(0)\alpha} - i\overline{\psi}^{(0)}\not{D}^{(0)}\psi^{(0)} - \sum_\beta \varepsilon^{(0)\beta}u^{(0)\beta} \end{aligned} \tag{16.1}$$

where

$$D_\mu^{(0)\alpha\beta} = \delta^{\alpha\beta}\partial_\mu + g_0 T^{\gamma\alpha\beta}\phi_\mu^{(0)\gamma}, \tag{16.2}$$

$$G_{\mu\nu}^{(0)\alpha} = \partial_\mu\phi_\nu^{(0)\alpha} - \partial_\nu\phi_\mu^{(0)\alpha} - g_0 f^{\alpha\beta\gamma}\phi_\mu^{(0)\beta}\phi_\nu^{(0)\gamma}. \tag{16.3}$$

The index α is an $SU_3\times SU_3$ index and will be split as follows: $\alpha\equiv(a,R)$ where a is an SU_3 index and $R=A$ for axial and $R=V$ for vector. The equations of motion for the unrenormalized fields follow directly from the above Lagrangian. One introduces

$$Z_0 = \frac{m^2}{m_0^2}, \qquad g = \frac{Z_0}{Z^{\frac{1}{2}}}g_0, \qquad \varepsilon^\alpha = \frac{g}{m^2}\; Y^{\frac{1}{2}}\varepsilon^{(0)\alpha} \tag{16.4}$$

and the renormalized fields

$$\phi_\mu^\alpha = Z^{-\frac{1}{2}}\phi^{(0)\alpha}, \tag{16.5}$$

$$G_{\mu\nu}^\alpha = Z^{-\frac{1}{2}}G_{\mu\nu}^{(0)\alpha}, \tag{16.6}$$

$$u^\alpha = Y^{-\frac{1}{2}}u^{(0)\alpha}. \tag{16.7}$$

The commutation relations are best written in compact form by introducing a unit timelike fourvector n_μ. From the field equations and from the canonical commutation relations one finds

$$\begin{aligned} & \delta(nx)\,[\phi_\mu^\alpha(x),\phi_\nu^\beta(0)] \\ & = i\frac{g}{m^2}f^{\alpha\beta\gamma}\left\{\frac{n_\mu}{n^2}\phi_\nu^\gamma(0) + \frac{n_\nu}{n^2}\phi_\mu^\gamma(0) - \frac{n_\mu n_\nu}{n^4}\,n\cdot\phi^\gamma(0)\right\}\delta^4(x) \\ & \quad + \frac{i}{m^2}\frac{Z_0}{Z}\delta^{\alpha\beta}\frac{\mathrm{d}}{\mathrm{d}n_\lambda}\left(\frac{n_\mu n_\nu}{n^2}\right)\partial^\lambda\delta^4(x), \end{aligned} \tag{16.8}$$

$$\begin{aligned} & \delta(nx)\,[G_{\mu\nu}^\alpha(x),\phi_\varrho^\beta(0)] \\ & = i\frac{g}{m^2}f^{\alpha\beta\gamma}\frac{n_\varrho}{n^2}G_{\mu\nu}^\gamma(0) + \frac{i}{Z}\delta^{\alpha\beta}\left(\frac{n_\mu}{n^2}g_{\nu\varrho} - \frac{n_\nu}{n^2}g_{\mu\varrho}\right)\delta^{(4)}(x), \end{aligned} \tag{16.9}$$

$$\delta(nx)\,[u^\alpha(x),\phi_\varrho^\beta(0)] = i\frac{g}{m^2}T^{\alpha\beta\gamma}\frac{n_\varrho}{n^2}u^\gamma\delta^{(4)}(x), \tag{16.10}$$

$$\delta(nx)\,[\Pi_\mu^\alpha(x),\phi_\varrho^\beta(0)] = iT^{\beta\alpha\gamma}\frac{n_\varrho}{n^2}\Pi^\gamma(0)\,\delta^{(4)}(x). \tag{16.11}$$

In Eq. (16.9) we have introduced

$$\bar{g}_{\mu\nu} = g_{\mu\nu} - \frac{n_\mu n_\nu}{n^2}. \tag{16.12}$$

The fields $\Pi^\alpha_\mu(x)$ are (for $\mu=0$) the conjugate momenta of $u^\alpha(x)$. The Fourier transform of the T^*-product of s local operators $A_j(x_j)$ ($j=1,2,\ldots s$) has a formal asymptotic expansion of the following type:

$$\int (\Pi d^4 x_j)\, e^{-i\Sigma q_j x_j}\, T^*\{\Pi A_j(x_j)\} \to \sum_J C^J(q_1,\ldots q_s). \tag{16.13}$$

In Eq. (16.13) the limit is for $|p_i|^2 \to \infty$ ($i=1,2,\ldots s-1$). The expression for $C^J(q_1,\ldots q_s)$ is

$$\begin{aligned} C^J(q_1,\ldots q_s) = &\sum_{\{k_i\}} \sum_{\{m_i\}} \sum_{\{j_i\}} \sum_{P\{l_i\}} \delta\left[\sum_{i=1}^{s-1} (k_i - m_i) - J\right] \\ &\times \int d^4x e^{-i(\Sigma q_i)x} H(\{m_i\},\{k_i\},\{j_i\}) \left[\prod_{i=1}^{s-1} \left(p'_i \frac{\partial}{\partial p_i}\right)^{j_i}\right] \\ &\times S^{(k_1)}_{p_{l_1},p'_{l_1}}(\partial^{m_1}_{p_{l_1}} S^{(k_2)}_{p_{l_2},p'_{l_2}}(\ldots S^{(k_s-1)}_{p_{l_{s-1}},p'_{l_{s-1}}}(\partial^{m_{s-1}}_{p_{l_{s-1}}} A_{l_s}; A_{l_{s-1}});\ldots; A_{l_2}); A_{l_1}). \end{aligned} \tag{16.14}$$

In Eq. (16.14) $\sum_{\{k_i\}}$ indicates summation over $k_1, k_2, \ldots k_{s-1}$, $\sum_{\{m_i\}}$ summation over $m_1, m_2, \ldots m_{s-1}$, $\sum_{\{j_i\}}$ over $j_1, j_2, \ldots j_{s-1}$, $\sum_{P\{l_i\}}$ is over certain permutations of $(1,2,\ldots s)$; $(l_1,\ldots l_s)$ is a permutation of $(1,2,\ldots s)$; $H(\{m_i\},\{k_i\},\{j_i\})$ are numerical coefficients which are non-vanishing only when all the following conditions are satisfied

$$j_i m_i = 0\,, \quad k_i m_i = 0$$

for $i=1,2,\ldots s-1$; the momenta p_i and p'_i ($i=1,2,\ldots s-1$) are linear combinations of the q_i; the symbol $S^{(k)}_{p,p'}(A,B)$ has the following meaning:

$$S^{(k)}_{p,p'}(A,B) \equiv p'^{\mu_1} \ldots p'^{\mu_k} S^{(k)}_{\mu_1,\mu_2,\ldots\mu_k}(A,B;n)|_{n=p} \tag{16.15}$$

where $S^{(k)}_{\mu_1,\ldots\mu_k}(A,B;n)$ is defined from

$$\begin{aligned} &\delta(n(x-y))\,[A(x),B(y)] \\ &\quad = \sum_{j=0}^{N} i^k S^{(k)}_{\mu_1,\ldots\mu_k}(A,B;n)\, \bar{\partial}^{\mu_1}_n \ldots \bar{\partial}^{\mu_k}_n \delta^{(4)}(x-y) \end{aligned} \tag{16.16}$$

with

$$\bar{\partial}^\mu_n = \partial^\mu - \frac{n\partial}{n^2} n^\mu \tag{16.17}$$

In Eq. (16.14) we have introduced

$$\partial_p^m \equiv \left(-i\,\frac{p^\mu \partial_\mu}{p^2}\right)^m .$$

The expansion in Eqs. (16.13), (16.14), is derived in Ref. [19] and [20]. In particular for $A_i = \phi_{\mu_i}^{a_i}$ the calculation of the asymptotic expansion of the Fourier transform of the T^*-product reduces to the calculation of expressions of the kind

$$S^{(k_1)}\left(\partial_{p_1}^{m_1} S^{(k_2)}\left(\ldots S^{(k_s-1)}\left(\partial_{p_{s-1}}^{m_{s-1}}\phi;\phi;p_{s-1}\right);\ \ldots;\phi;p_2\right);\phi;p_1\right) \tag{16.18}$$

in an abbreviated notation where tensor indices are omitted. It will be convenient to introduce operators $\mathscr{S}_p^{(k,m)}$ defined on the algebra of local operators A by

$$\mathscr{S}_p^{(k,m)} \cdot A \equiv S^{(k)}\left(\partial_p^m A;\phi;p\right). \tag{16.19}$$

The expression in Eq. (16.18) can then be written as

$$\prod_{j=1}^{j=s-1}{}^{\text{ord}}\ \mathscr{S}_p^{(k_j;m_j)} \cdot \phi .$$

Let us call $\mathscr{A}(x)$ the associative algebra containing the unit element generated from the canonical variables and their spatial derivatives. The field equations insure us that also ϕ_0^a and the time derivatives of the canonical variables and of ϕ_0 belong to $\mathscr{A}(x)$. Thus $\mathscr{A}(x)$ is (formally) closed under derivation. From the canonical commutators it follows that the operators S appearing as coefficients of the Schwinger terms in Eq. (16.16) belong to $\mathscr{A}(y)$ for $A \in \mathscr{A}(x)$ and $B \in \mathscr{A}(y)$. We then introduce the following definition:

Def.: Let us consider the operators

$$\prod_{j=1}^{j=s}{}^{\text{ord}}\ \mathscr{S}_{n_j}^{(k_j;0)} \cdot A \equiv S_{\{n_j\}}^{(K,\{k_j\};s)} \tag{16.20}$$

where $A \in \mathscr{A}(x)$, $n_j^2 > 0$, $j = 1, \ldots s$, and $K = \Sigma k_j$. The lowest non-negative value of K, to be called $\bar{K}$, such that $S_{\{n_j\}}^{(\bar{K},\{k_j\},\ s)} \equiv 0$ holds identically with respect to $\{n_j\}$ for each partition $(K, \{k_j\}; s)$ of $K > \bar{K}$, will be called the "rank" of A.

One verifies that the rank of a sum is non larger than the highest of the ranks in the sum; and, furthermore, the rank of a product is not larger that the sum of the ranks of the factors. The rank of the tensors $\bar{\partial}^{m-1}\phi$,

$\bar{\partial}^m G$, $\bar{\partial}^m u$, $\bar{\partial}^m \Pi$, $\bar{\partial}^m \psi$, and $\bar{\partial}^m \bar{\psi}$ is m. We also introduce the following definitions:

Def.: We call $V^{(k)}$ the subspace of $\mathscr{A}$ generated from the elements of $\mathscr{A}$ of rank not larger than k for $k \geqq 0$, for $k<0$ we put $V^{(k)}=0$.

Def.: We call $V(A \oplus B \oplus \cdots)$ the space generated from $A, B, \ldots$ all belonging to $\mathscr{A}$.

Def.: We write $A \overset{(k)}{\approx} B$ when $(A-B) \in V^{(k-1)}$ for $A, B \in \mathscr{A}(x)$.

Def.: We write $A \overset{(k)}{\in} V(B \oplus C \oplus \cdots)$ if $X \in V(B \oplus C \ldots)$ exists such that $A \overset{(k)}{\approx} X$.

One can show by induction the following property.

(I) The tensors $\partial^m \phi$, $\partial^m G$, $\partial^m \Pi$, $\partial^m u$ (∂^m is an abbreviation for $\partial_{\mu_1} \partial_{\mu_2} \ldots \partial_{\mu_m}$) are of rank $m+1$, m, m and m respectively and, neglecting terms of rank less than $m-1$, m, m, m respectively, they can be expressed in terms of $\bar{\partial}_n^k \phi$, $\bar{\partial}_n^k G$, $\bar{\partial}_n^k \Pi$, and u ($k=0, \ldots, m$; n_μ timelike) in the following form

$$\partial^m \phi \overset{(m-1)}{\in} V\left\{\frac{1}{\eta} P_n^{(m+1)}(\eta \phi) \oplus \ldots P_n^{(m-j-k-1)}(\eta \phi)\, \bar{\partial}^j G\, P_n^{(k)}(\eta \phi) \ldots \right.$$
$$\left. \oplus \ldots P_n^{(m-j-1)}(\eta \phi)\, u P_n^{(j)}(\eta \phi) \ldots \oplus \frac{1}{Z} P_n^{(m-1)}(\eta \phi)\right\}. \tag{16.21}$$

In Eq. (16.21) we have introduced $\eta = Z/Z_0$; the index j runs from $j=0$ to $j=m-1$, and k from $k=0$ to $k=m-j-1$. We have defined: for $K<0$, $P_n^{(k)} \equiv 0$; $P_n^{(0)}(\eta \phi) \equiv$ c-number; for $K \geqq 1$ $P_n^{(k)}(\eta \phi)$ is an element of $V\left\{\ldots \eta^j \bar{\partial}_n^{k_1} \phi \bar{\partial}_n^{k_2} \phi \ldots \bar{\partial}_n^{k_j} \phi\, \delta\left(\sum_{i=1}^{j} k_i + j - k\right) \ldots\right\}$ when $1 \leqq j \leqq K$. Relations similar to (16.21) hold for $\partial^m G$, $\partial^m u$, and $\partial^m \Pi$.

One also obtains:

(II) For each $A \in \mathscr{A}(x)$ and such that it can be expressed in terms of the operators $\bar{\partial}^k \phi$, $\bar{\partial}^k G$, $\bar{\partial}^k \Pi$, and u ($k=0, 1, \ldots$) only, the following relation holds

$$\mathscr{S}_n^{(k,0)} \cdot A \in V\left\{G \frac{\partial A}{\partial(\bar{\partial}^k G)} \oplus \phi \frac{\partial A}{\partial(\bar{\partial}^k \phi)} \oplus \Pi \frac{\partial A}{\partial(\bar{\partial}^k \Pi)} \oplus \delta_{k,0} u \frac{\partial A}{\partial u} \right.$$
$$\left. \oplus \frac{1}{Z} \frac{\partial A}{\partial(\bar{\partial}^k G)} \oplus \frac{1}{\eta} \frac{\partial A}{\partial(\bar{\partial}^{k-1} \phi)} \theta(k-1)\right\} \tag{16.22}$$

where $\delta_{k,0}$ is the Kronecker delta and $\theta(k)=1$ for $k \geqq 0$, $\theta(k)=0$ for $k<0$.

From I and II one finds:

$$\begin{aligned}\mathscr{S}_n^{(k,m)}\phi \overset{(m-k-2)}{\tilde{\in}} V\Big\{\frac{1}{\eta} P_n^{(m-k+1)}(\eta\phi)\ldots \\ \oplus P_n^{(m-j-k-l-1)}(\eta\phi)\,\bar{\partial}_n^j G P_n^{(l)}(\eta\phi)\ldots \\ \oplus P_n^{(m-k-j-1)}(\eta\phi)\,u P_n^{(j)}(\eta\phi)\ldots \\ \oplus \frac{1}{Z} P_n^{(m-k-1)}(\eta\phi)\Big\} = V_n^{(k,m)}\end{aligned} \tag{16.23}$$

where j runs form $j=0$ to $j=m-k-1$ and l from $l=0$ to $l=m-k-j-1$.

(III) More generally one can show that

$$\mathscr{S}_{n'}^{(k,m)}\phi \overset{m-k-2}{\tilde{\in}} V_n^{(k,m)} \tag{16.24}$$

for n_μ and n'_μ both time-like.

(IV) Eq. (16.24) can be iterated giving

$$\begin{aligned}\prod_{j=1}^{s}{}^{\mathrm{ord}}\, \mathscr{S}_{n_j}^{(k_j,m_j)}\phi \overset{(M_s-K_s-2)}{\tilde{\in}} V\Big\{\frac{1}{\eta} P_{n_1}^{(M_s-K_s+1)}(\eta\phi)\,\Pi_i\theta(M_i-K_i+1)\ldots \\ \oplus\ldots \sum_{j=0}^{M_s-K_s-l-1} P_{n_1}^{(M_s-K_s-l-j-1)}(\eta\phi)\,\bar{\partial}_{n_1}^l G\, P_{n_1}^{(j)}(\eta\phi)\,\Pi_i\theta(M_i-K_i-1)\ldots \\ \oplus \sum_{j=0}^{M_s-K_s-1} P_{n_1}^{(M_s-K_s-j-1)}(\eta\phi)\,u\, P_{n_1}^{(j)}(\eta\phi)\,\Pi_i\theta(M_i-K_i-1) \\ \oplus\frac{1}{Z} P_{n_1}^{(M_s-K_s-1)}(\eta\phi)\,\Pi_i\theta(M_i-K_i-1)\Big\}\end{aligned} \tag{16.25}$$

where

$$0 \leqq l \leqq M_s-K_s-1,\ M_j=\sum_{i=j}^{i=s} m_i, \quad \text{and} \quad K_j=\sum_{i=j}^{i=s} k_i\ (M_s=M, K_s=K).$$

From the property (IV) one concludes that

$$\begin{aligned} &C^J(q_1\ldots q_s)=0 \qquad \text{for } J>1 \\ &C^{(0)}(q_1\ldots q_s)\in V(\phi), \\ &C^1(q_1\ldots q_s)\propto\frac{1}{\eta} \\ &C^{-1}(q_1\ldots q_s)\in V\left(\frac{1}{Z}\oplus u\oplus G\oplus\partial\phi\oplus\eta\phi\phi\right). \end{aligned} \tag{16.26}$$

For truncated products and going to the limit of ECFA one has

$$C^J(q_1 \ldots q_s)_T = 0 \quad \text{for } J > 0\,,$$
$$C^0(q_1 \ldots q_s)_T \in V(\phi)\,, \tag{16.27}$$
$$C^{-1}(q_1 \ldots q_s)_T \in V(u \oplus \partial\phi)\,.$$

By a detailed analysis of the field equations and commutators one concludes that the u's appearing in $C^{-1}(q_1 \ldots q_s)_T$ are in fact originated from commutators of the kind

$$[Q, [Q, [\ldots, [Q, \partial^\mu \phi_\mu] \ldots]]]\,.$$

Before concluding we may comment on the very simple proof given by *Iliopoulos* [21] of the results contained in this section. The simplicity comes from an explicit assumption in that paper, that of keeping only terms proportional to current divergences. The remaining terms are just the cause of the complication here and they cannot be neglected unless the commutator algebra is such as to allow for their neglection (like here with CFA). It is clear why the proof by *Iliopoulos* goes through even after neglecting such terms: the reason is that there exists a consistent scheme where such terms, as shown here, can be neglected.

V. Further Speculations

17. An Alternative Approach

Here instead of following the renormalization approach we consider the possibility that the summed up leading weak divergences can be treated as finite and so counterterms are added to the original Lagrangian.

We take

$$\mathscr{L} = \overline{\psi}_0[-i\partial\!\!\!/ + M_0]\,\psi_0 + g_0\overline{\psi}_0\gamma_\mu W^\mu P_+ \Lambda(\theta_0)\,\psi_0 + \text{h.c.} \tag{17.1}$$

where

$$P_\pm = \tfrac{1}{2}(1 \pm i\gamma_5)\,, \tag{17.2}$$

$$\Lambda(\theta_0) = \tfrac{1}{2}(\lambda_1 + i\lambda_2)\cos\theta_0 + \tfrac{1}{2}(\lambda_4 + i\lambda_5)\sin\theta_0\,, \tag{17.3}$$

$$M_0 = \begin{bmatrix} p_0 & & \\ & n_0 & \\ & & \lambda_0 \end{bmatrix}. \tag{17.4}$$

We perform a wave-function renormalization

$$\psi_0 = (Y_+ P_+ + Y_- P_-)\,\psi \tag{17.5}$$

and rewrite $\mathscr{L}$ as follows

$$\mathscr{L} = \mathscr{L}_0 + \mathscr{L}_I\,, \tag{17.6}$$

$$\mathscr{L}_0 = \overline{\psi}(-i\partial\!\!\!/ + M)\,\psi\,, \tag{17.7}$$

$$M = \begin{bmatrix} p & & \\ & n & \\ & & \lambda \end{bmatrix}, \tag{17.8}$$

$$\mathscr{L}_I = g\overline{\psi}\gamma_\mu W^\mu \Lambda(\theta)\, P_+ \psi + C + \text{h.c.} \tag{17.9}$$

$$\begin{aligned} C = \overline{\psi}\{&(-i\partial\!\!\!/)\,(Y_+^\dagger Y_+ P_+ + Y_-^\dagger Y_- P_- - 1) \\ &+ (P_+ Y_+^\dagger M_0 Y_- + P_- Y_-^\dagger M_0 Y_+ - M)\}\,\psi \\ &+ \overline{\psi}\gamma_\mu W^\mu P_+ (g_0 Y_+^\dagger \Lambda(\theta_0)\, Y_+ - g\Lambda(\theta))\,\psi\,. \end{aligned} \tag{17.10}$$

For the leading divergences we have a Ward identity relating, as usual, vertex and propagator. In fact the term corresponding to the divergence of the current is not a leading divergent term (because leading divergences correspond to neglecting the internal masses, and such a term would thus be more convergent). We therefore require

$$g_0 Y_+^\dagger \Lambda(\theta_0)\, Y_+ - g\Lambda(\theta) = 0 \tag{17.11}$$

(there is no need for vertex counterterms of different tensor character).

Next we require

$$S_F^{-1}(p) = p\!\!\!/ - M\,. \tag{17.12}$$

Now $S_F^{-1}(p)$ calculated from (17.6) and (17.11) is

$$S_F^{-1}(p) = p\!\!\!/ - M - \Sigma(p) + C(p) \tag{17.13}$$

where

$$\Sigma(p) = -F p\!\!\!/ P_+ \Lambda(\theta)\,, \tag{17.14}$$

$$\Lambda(\theta) = \begin{bmatrix} 1 & 0 & 0 \\ 0 & \cos^2\theta & \sin\theta\cos\theta \\ 0 & \sin\theta\cos\theta & \sin^2\theta \end{bmatrix}. \tag{17.15}$$

From (17.12) and (17.13) one has

$$Y_-^\dagger Y_- = 1\,, \tag{17.16}$$

$$Y_+^\dagger Y_+ = 1 - F\,\Lambda(\theta)\,, \tag{17.17}$$

$$Y_-^\dagger M_0 Y_+ = M \quad \text{(and the hermitian conjugate equation } Y_+^\dagger M_0 Y_- = M)\,. \tag{17.18}$$

If Y_+, Y_- exists satisfying (17.11), (17.16), (17.17), (17.18) one has succeded in reducing the original Lagrangian, Eq. (17.1), to one where the leading weak divergences are compensated but the "physical" quarks have now physical masses as given from the matrix M and are coupled with a different Cabibbo angle.

The construction of Y_- and Y_+ and the relations between unrenormalized and renormalized quantities can be obtained by algebraic manipulations. We sketch the argument. Let us introduce (for arbitrary ϕ) the rotation matrix

$$R(\phi)=\begin{bmatrix}1 & 0 & 0\\ 0 & \cos\phi & \sin\phi\\ 0 & -\sin\phi & \cos\phi\end{bmatrix} \tag{17.19}$$

which satisfies

$$R(\phi_1+\phi_2)=R(\phi_1)\,R(\phi_2)\,. \tag{17.20}$$

$A(\theta)$, defined in Eq. (17.15), can be written as

$$A(\theta)=R(-\theta)\,ER(\theta) \tag{17.21}$$

with

$$E=\begin{bmatrix}1 & 0 & 0\\ 0 & 1 & 0\\ 0 & 0 & 0\end{bmatrix}. \tag{17.22}$$

Also

$$A(\theta)=R(-\theta)\,E'R(\theta) \tag{17.23}$$

with

$$E'=\begin{bmatrix}0 & 1 & 0\\ 0 & 0 & 0\\ 0 & 0 & 0\end{bmatrix}. \tag{17.24}$$

The representations (17.21) and (17.23), in conjunction with Eq. (17.20) will considerably simplify the discussion.

We have

$$1-FA(\theta)=R(-\theta)\,(1-FE)\,R(\theta)\,, \tag{17.25}$$

$$\det\,(1-FA(\theta))=\det\,(1-FE)=(1-F)^2\,. \tag{17.26}$$

It is essential that $A(\theta)$ is a projection operator. This fact is evident from Eq. (17.21)

$$A^2(\theta)=R(-\theta)\,E^2R(\theta)=R(-\theta)\,ER(\theta)=A(\theta) \tag{17.27}$$

since

$$E^2=E\,. \tag{17.28}$$

Eq. (17.17) can be written in the form

$$U^{\dagger} U = 1 \tag{17.29}$$

where

$$U = Y_+ [1 - hA(\theta)], \tag{17.30}$$

$$h = 1 + \frac{1}{\sqrt{1-F}}. \tag{17.31}$$

In fact inserting Eq. (17.30) into Eq. (17.29) one has

$$\begin{aligned} &[1 - hA(\theta)]\, Y_+^{\dagger} Y_+ [1 - hA(\theta)] = 1, \\ &Y_+^{\dagger} Y_+ = [1 - hA(\theta)]^{-2}. \end{aligned} \tag{17.32}$$

The inverse of $[1-hA(\theta)]$ is $[1-\bar{h}A(\theta)]$ with

$$\bar{h} = \frac{h}{h-1} \tag{17.33}$$

as can be verified immediately (see Eq. (17.27));

$$(1 - \bar{h}A(\theta))\,(1 - hA(\theta)) = 1 - (h + \bar{h} - \bar{h}h)\, A(\theta) = 1 \tag{17.34}$$

since one has

$$h + \bar{h} - \bar{h}h = h + \bar{h}(1-h) = 0.$$

Thus Eq. (17.17) only says that

$$Y_+ = U(1 - \bar{h}A(\theta)) \tag{17.35}$$

where U is unitary. Eq. (17.16) says that Y_- is unitary. It will be convenient to calculate M^2 from Eq. (17.18)

$$\begin{aligned} M^2 &= Y_+^{\dagger} M_0 Y_- Y_-^{\dagger} M_0 Y_+ = Y_+^{\dagger} M_0^2 Y_+ \\ &= M^2 = [1 - \bar{h}A(\theta)]\, U^{\dagger} M_0^2 U [1 - \bar{h}A(\theta)], \\ &[1 - hA(\theta)]\, M^2 [1 - hA(\theta)] = U^{\dagger} M_0^2 U. \end{aligned} \tag{17.36}$$

Similarly, from Eq. (17.11),

$$g_0 Y_+^{\dagger} \Lambda(\theta_0)\, Y_+ = g_0 [1 - \bar{h}A(\theta)]\, U^{\dagger} \Lambda(\theta_0)\, U [1 - \bar{h}A(\theta)] = g\Lambda(\theta)$$

or

$$U^{\dagger} \Lambda(\theta_0)\, U = \frac{g}{g_0} [1 - hA(\theta)]\, \Lambda(\theta)\, [1 - hA(\theta)]. \tag{17.37}$$

Eq. (17.37) can be simplified using the properties

$$A(\theta)\, \Lambda(\theta) = \Lambda(\theta), \tag{17.38}$$

$$\Lambda(\theta)\, A(\theta) = \Lambda(\theta). \tag{17.38'}$$

These equations are easily verified:

$$A(\theta)\,\Lambda(\theta) = R(-\theta)\,EE'R(\theta)\,,$$

$$\Lambda(\theta)\,A(\theta) = R(-\theta)\,E'ER(\theta)$$

but

$$EE' = E'E = E'\,. \tag{17.39}$$

Thus

$$[1 - hA(\theta)]\,\Lambda(\theta)\,[1 - hA(\theta)] = [1 - h]\,\Lambda(\theta)\,[1 - hA(\theta)] = (1-h)^2\,\Lambda(\theta)$$

giving for Eq. (17.37)

$$U^\dagger \Lambda(\theta_0)\,U = \frac{g}{g_0}(1-h)^2\,\Lambda(\theta)\,. \tag{17.40}$$

From Eq. (17.40) we get

$$U^\dagger \Lambda(\theta_0)\,U\,U^\dagger \Lambda^\dagger(\theta_0)\,U = \left[\frac{g}{g_0}(1-h)^2\right]^2 \Lambda(\theta)\,\Lambda^\dagger(\theta)\,.$$

Taking the trace

$$\mathrm{Tr}\,[\Lambda(\theta_0)\,\Lambda^\dagger(\theta_0)] = \left[\frac{g}{g_0}(1-h)^2\right]^2 \mathrm{Tr}\,[\Lambda(\theta)\,\Lambda^\dagger(\theta)]\,.$$

Noting that, for any ϕ,

$$\mathrm{Tr}\,[\Lambda(\phi)\,\Lambda^\dagger(\phi)] = \mathrm{Tr}\,[R(-\phi)\,E'R(\phi)\,R(-\phi)\,E'^\dagger R(\phi)] = \mathrm{Tr}\,[E'E'^\dagger]$$

one has

$$\left[\frac{g}{g_0}(1-h)^2\right]^2 = 1\,. \tag{17.41}$$

Choosing the sign + in extracting the square root of Eq. (17.41) (i.e. g_0 of same sign as g) we have finally

$$U^\dagger \Lambda(\theta_0)\,U = \Lambda(\theta)\,. \tag{17.42}$$

Let us put

$$U = R(\theta - \theta_0)\,V \tag{17.43}$$

thus requiring

$$V = R(\theta_0 - \theta)\,U\,.$$

One has

$$VV^\dagger = R(\theta - \theta_0)\,UU^\dagger R(\theta - \theta_0) = 1 \tag{17.44}$$

that is, V is unitary.

Inserting into Eq. (17.42), we have

$$V^{\dagger} R(\theta_0-\theta)\, \Lambda(\theta_0)\, R(\theta-\theta_0)\, V=\Lambda(\theta)\,.$$

We note the property, for general ϕ and χ,

$$R(-\phi)\,\Lambda(\chi)\,R(\phi)=R(-\phi)\,R(-\chi)\,E' R(\chi)\,R(\phi)=R(-\phi-\chi)\,E' R(\phi+\chi)\,,$$
$$R(-\phi)\,\Lambda(\chi)\,R(\phi)=\Lambda(\phi+\chi)\,. \qquad (17.45)$$

We thus have

$$V^{\dagger}\Lambda(\theta)\,V=\Lambda(\theta)\,, \qquad V V^{\dagger}=1\,. \qquad (17.46)$$

It follows from Eqs. (17.46) that

$$V=\mathrm{e}^{i\xi}1 \qquad (17.47)$$

where ξ is an unessential phase (that we shall choose to be zero). The proof is straightforward.

We have thus found that

$$Y_+=U[1-\bar{h}A(\theta)]=R(\theta-\theta_0)\,[1-\bar{h}A(\theta)]\,. \qquad (17.48)$$

Inserting $U=R(\theta-\theta_0)$ into Eq. (17.36) we have

$$[1-hA(\theta)]\,M^2[1-hA(\theta)]=R(\theta_0-\theta)\,M_0^2 R(\theta-\theta_0) \qquad (17.49)$$

and the inverse equation

$$[1-\bar{h}A(\theta)]\,M^{-2}[1-\bar{h}A(\theta)]=R(\theta_0-\theta)\,M_0^{-2}R(\theta-\theta_0)\,. \qquad (17.50)$$

Let us now use Eq. (17.18) in the form

$$Y_-^{\dagger}=M\,Y_+^{-1}\,M_0^{-1}$$

and, from Eqs. (17.16) and (17.48)

$$Y_-=(Y_-^{\dagger})^{-1}=M_0\,Y_+\,M^{-1}=M_0 R(\theta-\theta_0)\,[1-\bar{h}A)(\theta)]\,M^{-1}\,.$$

It follows

$$Y_-\,Y_-^T=M_0\,Y_+\,M^{-1}M M^{-1}\,Y_+^T M_0$$
$$=M_0 R(\theta-\theta_0)\,[1-\bar{h}A(\theta)]\,M^{-2}[1-\bar{h}A(\theta)]\,R(\theta_0-\theta)\,M_0$$

since $A^T(\theta)=A(\theta)$. Inserting Eq. (17.50) we have

$$Y_-\,Y_-^T=M_0 R(\theta-\theta_0)\,R(\theta_0-\theta)\,M_0^{-2}R(\theta-\theta_0)\,R(\theta_0-\theta)\,M_0$$

or

$$Y_-\,Y_-^T=1\,. \qquad (17.51)$$

Eqs. (17.51) and (17.16) give

$$Y_- = Y_-^* = R(\beta) \tag{17.52}$$

where β is a well-determined angle.

To determine the relations between physical and unphysical quantities we proceed as follows. From Eq. (17.49) we have

$$p_0^2 = x^2 p^2 . \tag{17.53}$$

We take the trace of Eq. (17.49) and obtain, using Eq. (17.53)

$$\begin{aligned}\mathrm{Tr}\, M_0^2 &= \mathrm{Tr}\,\{[1 - hA(\theta)]\, M^2 [1 - hA(\theta)]\} = \mathrm{Tr}\,\{M^2 [1 - hA(\theta)]^2\} \\ &= \mathrm{Tr}\,\{M^2 R(-\theta)\,[1 - hE]^2\, R(\theta)\} = \mathrm{Tr}\,\{R(\theta)\, M^2 R(-\theta)\,(1 - hE)^2\}\end{aligned}$$

or

$$n_0^2 + \lambda_0^2 = x^2 (n^2 \cos^2\theta + \lambda^2 \sin^2\theta) + n^2 \sin^2\theta + \lambda^2 \cos^2\theta \tag{17.54}$$

with

$$x = 1 - h . \tag{17.55}$$

Taking the determinant of Eq. (17.49), we obtain

$$\det M_0^2 = \det M^2 \cdot \det (1 - hA)^2 = x^2 \det M^2 \tag{17.56}$$

or, using Eq. (17.53)

$$n_0^2 \lambda_0^2 = x^2 n^2 \lambda^2 \tag{17.57}$$

Again, from Eq. (17.49),

$$R(\theta)\,[1 - hA(\theta)]\, M^2 [1 - hA(\theta)]\, R(-\theta) = R(\theta_0)\, M_0 R(-\theta_0)$$

or, using Eq. (17.21),

$$\begin{aligned}&(1 - hE)\, R(\theta)\, M^2 R(-\theta)\,(1 - hE) = R(\theta_0)\, M_0^2 R(-\theta_0)\,, \\ &R(\theta)\, M^2 R(-\theta) = (1 - hE)^{-1} R(\theta_0)\, M_0^2 R(-\theta_0)\,(1 - hE)^{-1} .\end{aligned}$$

Taking the trace, one has

$$\mathrm{Tr}\, M^2 = \mathrm{Tr}\,\{R(\theta_0)\, M_0^2 R(-\theta_0)\,(1 - hE)^{-2}\} .$$

Recalling Eq. (17.53) one obtains

$$n^2 + \lambda^2 = n_0^2 \sin^2\theta_0 + \lambda_0^2 \cos^2\theta_0 + \frac{1}{x^2}(n_0^2 \cos^2\theta_0 + \lambda_0^2 \sin^2\theta_0) . \tag{17.58}$$

Eqs. (17.53), (17.54), (17.57), and (17.58) provide the full set of equations relating physical to unphysical quantities. Thus the full set of equations is

$$x^2 = \frac{p_0^2}{p^2} = \frac{n_0^2 \lambda_0^2}{n^2 \lambda^2}, \tag{17.59}$$

$$n_0^2 + \lambda_0^2 = \frac{n_0^2 \lambda_0^2}{n^2 \lambda^2} (n^2 c^2 + \lambda^2 s^2) + (n^2 s^2 + \lambda^2 c^2), \tag{17.60}$$

$$n^2 + \lambda^2 = \frac{n^2 \lambda^2}{n_0^2 \lambda_0^2} (n_0^2 c_0^2 + \lambda_0^2 s_0^2) + (n_0^2 s_0^2 + \lambda_0^2 c_0^2) \tag{17.61}$$

with the abbreviations $c = \cos\theta$, $s = \sin\theta$, $c_0 = \cos\theta_0$, $s_0 = \sin\theta_0$. An alternative form for the set of Eqs. (17.60) and (17.61) is

$$t^2 = \frac{n^2}{\lambda^2} \frac{\lambda_0^2 - \lambda^2}{n_0^2 - n^2} \frac{\lambda^2 - n_0^2}{\lambda_0^2 - n^2}, \tag{17.62}$$

$$\frac{t_0^2}{t^2} = \frac{\lambda^2}{n^2} \frac{n_0^2}{\lambda_0^2} \frac{(\lambda_0^2 - n^2)^2}{(\lambda^2 - n_0^2)^2}, \tag{17.63}$$

where $t = \operatorname{tg}\theta$ and $t_0 = \operatorname{tg}\theta_0$.

18. Discussion and Speculations

Let us first note that for $n_0^2 = n^2$ and $\lambda_0^2 = \lambda^2$ one would obtain $p_0^2 = p^2$ from (17.59) and both (17.60) and (17.61) would be identically satisfied for any θ and for any θ_0; (the same conclusion follows from $p_0^2 = p^2$ and $\lambda_0^2 = \lambda^2$; or $p_0^2 = p^2$ and $n_0^2 = n^2$). In the approach developed in the previous sections a counterterm proportional to a fixed linear combination of the operators u_0, u_6, and v_7 was introduced. No particular conclusions can be drawn from the set of Eqs. (17.59)–(17.61) since, assuming p, n, λ to be known, one still has four quantities p_0, n_0, λ_0 and θ_0 completely unknown, in addition to θ. [An inequality must be required in order to have $t^2 \geqq 0$, namely

$$\frac{\lambda_0^2 - \lambda^2}{\lambda_0^2 - n^2} \frac{n_0^2 - \lambda^2}{n_0^2 - n^2} \leqq 0, \tag{18.1}$$

which limits the values of λ_0^2 and n_0^2 to the shadowed regions of the figure.]

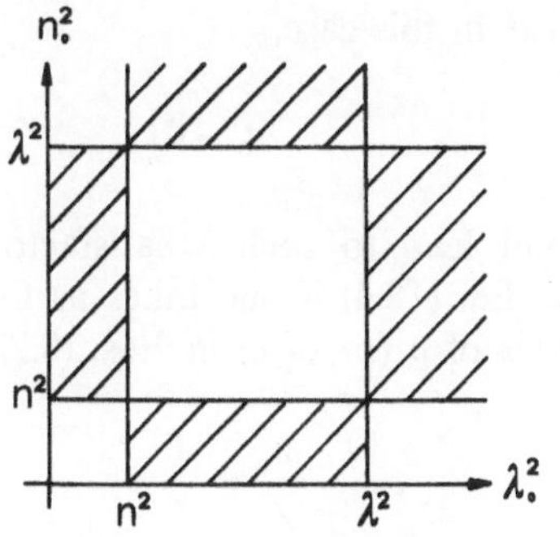

A counterterm proportional to u_0 would require the condition $M = M_0 + \Delta \cdot 1$, that is $p_0 = p + \Delta$, $n_0 = n + \Delta$, and $\lambda_0 = \lambda + \Delta$. We shall first impose only the conditions $n_0 = n + \Delta$ and $\lambda_0 = \lambda + \Delta$, corresponding to a slightly more general counterterm. Inserting $n_0 = n + \Delta$ and $\lambda_0 = \lambda + \Delta$ one has

$$t^2 = \frac{n^2}{\lambda^2} \frac{2\lambda + \Delta}{2n + \Delta} \frac{(\lambda - n) - \Delta}{(\lambda - n) + \Delta}, \tag{18.2}$$

$$\frac{t_0^2}{t^2} = \left(\frac{\lambda}{n} \frac{n + \Delta}{\lambda + \Delta} \frac{(\lambda - n) + \Delta}{(\lambda - n) - \Delta} \right)^2. \tag{18.3}$$

The result $\mathrm{tg}^2\,\theta = n/\lambda$ (see Eq. (11.10)) is recovered in this approach in the limit $\Delta \to 0$. In this limit one has

$$t^2 \to \frac{n}{\lambda}, \tag{18.4}$$

$$\frac{t_0^2}{t^2} \to 1. \tag{18.5}$$

Eq. (18.4) is the desidered result and Eq. (18.5) is significant, suggesting $\theta_0 = \theta$.

Therefore small Δ will give a very satisfactory situation. Another interesting limit obtains if one instead requires $n_0 \to 0$ (i.e. $\Delta \to -n$). One has in this case $\lambda_0 \to \lambda - n$, and $p_0 \to 0$. The situation $n_0 = 0$, $p_0 = 0$ corresponds to an original chiral $SU_2 \times SU_2$ before adding weak interactions. For the angles one has

$$t^2 = \frac{n}{\lambda} \frac{2\lambda - n}{\lambda - 2n} \cong 2 \frac{n}{\lambda} \tag{18.6}$$

since $(n/\lambda) \ll 1$. Also, in this case,

$$t_0^2 \to 0. \tag{18.7}$$

Eq. (18.6) does not lead to such a satisfactory determination of the Cabibbo angle as Eq. (18.4) if one takes as final the determination of n/λ from the values of ϱ (or of c) in Eqs. (8.22) and (8.23). From Eqs. (10.3) one has

$$\frac{n}{\lambda} \cong \frac{1}{3}\varrho .$$

A factor $\sqrt{2}$ in the value of $\mathrm{tg}\,\theta$ would modify the two determinations of Eqs. (11.5) and (11.5′) into the values $\mathrm{tg}\,\theta = 0.27$ and $\mathrm{tg}\,\theta = 0.31$ respectively. Although numerically perhaps less satisfactory than the result $t^2 = n/\lambda$, the possibility in Eq. (18.6) still deserves however some attention, also in view of the quite recognized possibility that the values of ϱ (or of c) as deduced in Section 8, are subject to large errors (a 40% error in the value of ϱ would make Eq. (18.6) compatible with experiment). The attractive feature of the solution $n_0 \to 0$, $p_0 \to 0$ is that such a limit corresponds to an exact $SU_2 \times SU_2$ for the unphysical particles. Besides one has in the limit $\theta_0 = 0$ as indicated from Eq. (18.7). We note also that the alternative limit $\lambda_0 \to 0$ (i.e. $\Delta \to -\lambda$) gives $p_0 \to 0$, $n_0 \to n - \lambda$, leading again to Eq. (18.6) but in this case $\theta_0 \to \pm \pi/2$.

If we try the full set of conditions, $p_0 = p + \Delta$, $n_0 = n + \Delta$, $\lambda_0 = \lambda + \Delta$, the value of Δ is determined from the values of p, n, and λ. A solution is $\Delta = \dfrac{n\lambda}{p} - \lambda - n$ giving

$$t^2 = \frac{n^2}{\lambda^2}\,\frac{2\lambda(1+\frac{1}{2}\delta) - n}{\lambda\delta + n}\,\frac{\lambda(1-\delta)}{\lambda(1+\delta) - 2n} \tag{18.8}$$

where we have put

$$\frac{n}{p} = 1 + \delta .$$

It is interesting that for $\delta \to 0$ (i.e. exact SU_2 for the "physical" particles) one has $\Delta = -n$, thus recovering the solution in Eqs. (18.6) and (18.7). One would have in this case: $n_0 \to 0$, $p_0 \to 0$ (i.e. $SU_2 \times SU_2$ for the "unphysical particles), $\theta_0 \to 0$, and $t^2 \cong 2(n/\lambda)$. In the above discussion we have limited ourselves to considering some possibilities. Additional possibilities can be contemplated but we shall not try to make a complete discussion. The reader will recognize that the possibility $\Delta \to 0$, leading to Eqs. (18.4) and (18.5) is closely related to the approach of the previous sections (see Section 12) and the subsequent extension to higher orders whenever it happens that the summed up leading divergences give in fact a small contribution. In particular it practically coincides with the perturbative limit briefly examined at the beginning of Section 12. The possibility leading to Eqs. (18.6) and (18.7), i.e. $\Delta \to -n$, is rather different. It has the interesting feature of corresponding to a limit $p_0 \to 0$, $n_0 \to 0$, (i.e. $SU_2 \times SU_2$ for the unphysical particles) and $\theta_0 \to 0$. It leads however to a value for the Cabibbo angle less acceptable unless the determination of the parameter ϱ (or equivalently of c) from strong interactions is in error by some 40%. Finally the last of the possibilities considered, leading to Eq. (18.8) with $\delta \to 0$, appears also of interest in that no SU_2 breaking has been introduced neither for physical particles nor for unphysical particles. Practically, the SU_2 input for physical particles leads here to exact $SU_2 \times SU_2$ for unphysical particles and to $\theta_0 \to 0$.

Again the Cabibbo angle so determined is only of the right magnitude and one should rely on the possibility of errors in ϱ, as mentioned above, or just admit that the explicit neglection of any SU_2 breaking in this case is responsible for the unaccuracy. In conclusion we do not want to advocate the viewpoint that the discussion in this section stands on a better footing than the approach presented in the preceding sections, based on introducing a more complex counterterm. Both approaches cannot be claimed to be definite. We have tried here to present in an exhaustive way our present knowledge of the problem, as we have gained after one year of work on the subject, including what appear to us the most interesting possibilities, essentially in a critical way. Our feeling however is that the results obtained, and possibly Eq. (11.3), are correct and that a nearer understanding of the problem may sooner or later be gained by more work on the subject.

VI. Related Work

A large amount of work has appeared in the last two years related to the problems discussed in these lectures. In particular papers have appeared on the question of the origin of the Cabibbo angle, which are very closely connected to the approach we have presented here. We shall not attempt to produce a bibliography but merely quote the papers we have noted, occasionally adding some comments to indicate the content. We shall broadly distinguish three items: a) The problem of weak divergences; b) $SU_3 \times SU_3$-breaking (including spontaneous breaking); c) Origin of the Cabibbo angle.

a) Problem of Weak Divergences

A full bibliography on the problem of divergences of weak interactions would have to be very long. We shall refer to the recent papers by: *F. E. Low* [22], *N. Christ*, [23], *T. D. Lee* [12], *M. Gell-Mann, M. L. Goldberger, M. M. Kroll, F. E. Low* [24], *T. D. Lee* and *G. C. Wick* [10], *G. Segre* [25], *H. Terazawa* [26], *L. I. Li* and *G. Segre* [27], *T. Appelquist* and *G. Carlson* [13], *R. N. Mohapatra* and *J. Subba Rao* [28].

Recent methods based on special summation procedures (see *G. Efimov* [29], *E. S. Fradkin* [30], *R. Delbourgo, A. Salam*, and *J. Strathdee* [31], where additional references can be found) might turn to be useful for weak interactions.

Among previous works on weak divergences we note the peratization approach by *G. Feinberg* and *A. Pais* [32], and some particular calculations, such as for instance those of *B. L. Ioffe* and *E. P. Shabalin* [33] and of *R. N. Mohapatra, J. Subba Rao*, and *R. E. Marshak* [34].

b) $SU_3 \times SU_3$-Breaking

In addition to the papers by *S. Glashow* and *S. Weinberg* [7] and by *M. Gell-Mann, R. J. Oakes* and *B. Renner* [7], there are many recent contributions. We refer to the review given by *B. Renner* [7] at *Lund* [35].

On the subject of spontaneous breaking [36, 37] of $SU_3 \times SU_3$ we would like to mention the recent work by *L. Michel* and *L. Radicati* [38] (see also *G. F. Cicogna, F. Strocchi*, and *R. Vergara-Caffarelli* [39]).

Spontaneous breaking of SU_3 was treated, among others, by: *R. Broat* [40], *N. Cabibbo* [41], and *L. Michel* and *L. Radicati* [42].

c) Origin of the Cabibbo Angle

The approach by *Cabibbo* and *Maiani* [43] is essentially identical to the one presented here in the main idea of relating the angle θ to the weak self-masses and in its formal development. The physical interpretation is however different and rests on the requirement of a consistency condition to be satisfied by the different interaction and which can be satisfied only if additional u-spin invariant contributions are included. In a recent paper [44] *Oakes* has shown that the $\eta \rightarrow 3\pi$ amplitudes that one can calculate from the term u_3 which arises when $u_0 - \sqrt{2} u_8$ is rotated through 2θ about the seventh direction in SU_3 is in reasonable agreement with experiment. We would like to mention also the recent paper by *Tanaka* and *Tarjanne* [45] based on solving a set of linear equations for spontaneous breaking (see *N. Cabibbo* [41]). The modifications of the problem in the presence of neutral weak currents are considered, for special models, in the first paper in Ref. [2] and also by *C. H. Albright* [46].

References

1. *Cabibbo, N.:* Phys. Rev. Letters **10**, 531 (1963).
2. *Gatto, R., Sartori, G., Tonin, M.:* Phys. Letters **28 B**, 128 (1968).
   — — — Convergence conditions and Cabibbo angle. Preprint University of Padua, July 1969, presented at the International Conference on High Energy Physics, August 1969.
3. — — — Lett. Nuovo Cimento **1**, 1 (1969); 399 (1969); see also *Gatto, R.:* Proceedings Coral Gables.
4. *Bjorken, J. D.:* Phys. Rev. **148**, 1467 (1966).
5. *Johnson, K., Low, F.:* Progr. Theoret. Phys. Japan Suppl. **37—38**, 74 (1966).
6. *Gell-Mann, M.:* Physics **1**, 63 (1964).
7. *Glashow, S., Weinberg, S.:* Phys. Rev. Letters **20**, 22 (1968).
   Gell-Mann, M., Oakes, R. J., Renner, B.: Phys. Rev. **175**, 2195 (1968).
8. *Bouchiat, C., Iliopoulos, J., Prentki, J.:* Nuovo Cimento **A 56**, 1150 (1968).
9. *Mathur, V. S., Rao, J. S.:* Rochester Preprint (1969).
10. *Lee, T. D., Wick, G. C.:* Nucl. Phys. **B 9**, 209 (1969).
11. *Stückelberg, E. C. G.:* Helv. Phys. Acta **11**, 225 (1938); 299 (1938).
12. *Lee, T. D.:* Nuovo Cimento **A 59**, 579 (1969).
13. *Appelquist, T., Carlson, G.:* Slac Pub. 633, July (1969).
14. *Kroll, N., Lee, T. D., Zumino, B.:* Phys. Rev. **157**, 376 (1967).
15. *Gell-Mann, M., Zachariasen, F.:* Phys. Rev. **124**, 953 (1961).
16. *Lee, T. D., Weinberg, S., Zumino, B.:* Phys. Rev. Letters **18**, 1029 (1967).
17. *Ciccariello, S., Sartori, G., Tonin, M.:* Nuovo Cimento **A 55**, 847 (1968).
18. *Gatto, R.:* e^+e^- annihilation into hadrons (sum rules and asymptotic behaviour). Proceedings of the 1969 Daresbury Conference on High Energy Electrons and Photons, Liverpool 1969.
19. *Ciccariello, S., Sartori, G., Tonin, M.:* Nuovo Cimento **A 63**, 846 (1969).
20. *Vendramin, I.:* To be published.
21. *Iliopoulos, J.:* Nuovo Cimento **A 62**, 209 (1969); see also *Mohapatra, R. N., Olesen, P.:* Phys. Rev. **179**, 1417 (1969).
22. *Low, F. E.:* Comments Nucl. Particle Phys. **2**, 33 (1968).
23. *Christ, N.:* Phys. Rev. **176**, 3086 (1968).
24. *Gell-Mann, M., Goldberger, M. L., Kroll, N. M., Low, F. E.:* Phys. Rev. **179**, 1518 (1969).
25. *Segré, G.:* Phys. Rev. **181**, 1996 (1969).
26. *Terazawa, H.:* Phys. Rev. Letters **22**, 254 (1969).
27. *Li, L. F., Segré, G.:* Preprint, University of Pennsylvania 1969.
28. *Mohapatra, R. N., Subba Rao, J.:* Nuovo Cimento **A 65**, 153 (1970).
29. *Efimov, G.:* Soviet Phys. JETP **17**, 1417 (1963).
30. *Fradkin, E. S.:* Nuclear Phys. **49**, 624 (1963).
31. *Delbourgo, R., Salam, A., Strathdee, J.:* Preprint ICTP, Trieste 69, 17.
32. *Feinberg, G., Pais, A.:* Phys. Rev. **131**, 2724 (1963).
33. *Ioffe, B. L., Shabalin, E. P.:* Soviet J. Nucl. Phys. **6**, 828 (1967).
34. *Mohapatra, R. N., Subba Rao, J., Marshak, R. E.:* Phys. Rev. Letters **20**, 1081 (1968).
35. *Renner, B.:* Proceedings of the 1969 Lund Conference on Elementary Particles, Lund 1969.

36. *Nambu, Y., Jona-Lasinio, G.:* Phys. Rev. **122**, 345 (1961); **124**, 246 (1961).
37. *Goldstone, J., Salam, A., Weinberg, S.:* Phys. Rev. **127**, 965 (1962).
38. *Michel, L., Radicati, L.:* Preprint, Scuola Normale Superiore, Pisa, 1969.
39. *Cicogna, G. F., Strocchi, F., Vergara-Caffarelli, R.:* Phys. Rev. Letters **22**, 497 (1969).
40. *Brout, R.:* Nuovo Cimento **A 47**, 1252 (1967).
41. *Cabibbo, N.:* In: *Zichichi, A.* (Ed.): Hadrons and their Interactions. New York: Academic Press 1968.
42. *Michel, L., Radicati, L.:* Proceedings of the V Coral Gables Conference, edited by *A. Perlmutter, C. A. Hurst* and *B. Kursunoglu*. New York 1968.
43. *Cabibbo, N., Maiani, L.:* Phys. Letters **B 28**, 131, (1968) and preprint Istituto Superiore di Sanità, ISS 69/18 — May 1969.
44. *Oakes, R. J.:* Phys. Letters **B 30**, 262 (1969).
45. *Tanaka, T., Tarjanne, P.:* Phys. Rev. Letters **23**, 1137 (1969).
46. *Albright, C. H.:* CERN preprint TH 1066 (1969).

Prof. Dr. *R. Gatto*
Istituto di Fisica dell'Università, Padova
Istituto Nazionale di Fisica Nucleare, Sezione di Padova
Via F. Marzolo 8
I-35100 Padova